ROLE OF RESERVOIR OPERATION IN SUSTAINABLE WATER SUPPLY TO SUBAK IRRIGATION SCHEMES IN YEH HO RIVER BASIN

Mawiti Infantri Yekti

Thesis committee

Promotor
Prof. Dr E. Schultz
Emeritus Professor of Land and Water Development
UNESCO-IHE Institute for Water Education
Delft, the Netherlands

Co-promotors
Prof. Dr I Nyoman Norken
Department of Civil Engineering
Faculty of Engineering, Udayana University
Bali, Indonesia

Dr László Hayde
Chair group Land and Water Development
Department of Water Sciences and Engineering
UNESCO-IHE Institute for Water Education
Delft, the Netherlands

Other members
Prof. Dr R. Uijlenhoet, Wageningen University & Research
Prof. Dr N.C. van de Giesen, Delft University of Technology
Prof. Dr I Wayan Windia, Udayana University, Bali, Indonesia
Prof. Dr P. van der Zaag, UNESCO-IHE, Delft, the Netherlands

This research was conducted under the auspices of the SENSE Research School for Socio-Economic and Natural Sciences of the Environment

ROLE OF RESERVOIR OPERATION
IN SUSTAINABLE WATER SUPPLY
TO SUBAK IRRIGATION SCHEMES
IN YEH HO RIVER BASIN

Thesis

submitted in fulfilment of the requirements of
the Academic Board of Wageningen University and
the Academic Board of the UNESCO-IHE Institute for Water Education
for the degree of doctor
to be defended in public
on Thursday 1 June, 2017 at 11:30 a.m.
in Delft, the Netherlands

by

Mawiti Infantri Yekti
Born in Jember, Indonesia

CRC Press/Balkema is an imprint of the Taylor & Francis Group, an informa business

Published by
CRC Press/Balkema
PO Box 11320, 2301 EH Leiden, The Netherlands
e-mail: Pub.NL@taylorandfrancis.com
www.crcpress.com - www.taylorandfrancis.com

ISBN 978-1-138-06543-7 (Taylor & Francis Group)
ISBN 978-94-6343-083-8 (Wageningen University)
DOI: http://dx.doi.org/10.18174/404538

Table of contents

Table of contents v

Acknowledgements ix

1 Introduction 1

2 Background and objectives 7

    2.1 Profile of the region 7

    2.2 Water resources 11

        2.2.1 World water resources 11

        2.2.2 Water resources in Indonesia 13

        2.2.3 Water management in Indonesia 14

        2.2.4 Water resources in Bali 16

    2.3 Definition of Subak irrigation schemes 17

    2.4 Subak schemes 19

        2.4.1 Paddy terraces 19

        2.4.2 Subak irrigation system 21

    2.5 Subak cultivation area 23

    2.6 Previous studies on water management of Subak irrigation schemes 23

    2.7 Problem description 26

    2.8 Objectives 29

        2.8.1 Research questions 29

        2.8.2 Hypotheses 29

        2.8.3 Research objectives 31

3 Development of Subak irrigation schemes: learning from experiences of
ancient Subak schemes for participatory irrigation *system* management in Bali 33

    3.1 Introduction 33

    3.2 Method and discussion 36

        3.2.1 PIM in irrigation system operation and maintenance 39

        3.2.2 PIM with respect to socio-culture and economics of agriculture 45

        3.2.3 PIM in light of a religious community 51

    3.3 Result and conclusion 54

4   Subak in the south of Bali: discharge analysis for a system approach to river
     basin development with Subak irrigation schemes as a culture heritage 57
     4.1   Introduction 57
     4.2   Study of a river basin 57
     4.3   Managed flow approach in Yeh Ho River Basin 59
     4.4   Method and material.. 59
     4.5   Results and discussion... 62
     4.6   Conclusion... 73
5   Hydrology and hydraulic approaches: irrigation-drainage of Subak irrigation
     schemes, a farmer's perspective over a thousand years.. 77
     5.1   Introduction 77
     5.2   Methodology 78
            5.2.1 Observation of the water balance in a paddy terraces block................. 80
     5.3   Results and discussion... 80
     5.4   Conclusion... 87
6   Model simulations and optimisation technique 89
     6.1   Model categorization... 89
     6.2   Modelling of Subak schemes related to *Tri Hita Karana* philosophy 91
     6.3   Multiple purpose reservoir operation 92
     6.4   Scenario analysis in Subak schemes 92
     6.5   Simulations with the RIBASIM model 94
     6.6   Application aspects of the RIBASIM model 95
     6.7   Yeh Ho River system as input in the RIBASIM model............................... 97
7   Scenario analysis 101
     7.1   Hydrologic and hydraulic analysis.. 101
            7.1.1 Analysis of rainfall data: dependable rainfall and effective rainfall ... 102
            7.1.2 Streamflow analysis 104
            7.1.3 Potential evapotranspiration.. 104
            7.1.4 Reservoir water surface losses and gains 105
            7.1.5 Other reservoir losses and gains... 106
            7.1.6 Reservoir elevation/storage/area relationship 106

7.1.7 Flow routing in the reservoir and hydraulic profile of outlets 107

7.1.8 Evaluation of reservoir lifetime based on sedimentation.................... 107

7.2   Advanced irrigation node in RIBASIM 108

7.2.1 Schematization of the irrigated area... 109

7.2.2 Interactive graphical cropping plan editor 111

7.2.3 Simulation of a cropping plan.. 113

7.2.4 Soil moisture characteristics 114

7.2.5 Crop water requirement in a paddy terraces block............................ 118

7.2.6 Computation of command area water demand, actual field water

balance and effective irrigation water supply 122

7.3   Results of economic evaluation of storage allocation................................. 125

7.3.1 Pricing of paddy productivity.. 125

7.3.2 Pricing of domestic water.. 126

7.4   Scenario analysis, simulation and optimisation of Yeh Ho River Basin 129

7.4.1 Simulation of the first scenario 131

7.4.2 Simulation of the second scenario... 136

7.4.3 Simulation of the third scenario 140

7.4.4 Simulation of the fourth scenario.. 145

7.4.5 Simulation of the fifth scenario... 149

7.5   Summary of the simulation and optimisation of Yeh Ho River Basin 155

7.5.1 Utilisation of hydraulic structures... 155

7.5.2 Verification of the model 156

8   Evaluation 159

8.1   Recommendations for river basin development... 159

8.1.1 Telaga Tunjung Reservoir operation based on Subak cropping

patterns 160

8.1.2 Operation and maintenance of the Subak irrigation systems............. 161

9   References.. 163

APPENDICES

A. Abbreviations and acronyms 175

B. Symbols 177

C. Surface runoff analysis as inflow to the Telaga Tunjung Reservoir 179

D. Analysis of rainfall data 181

E. Reference evapotranspiration 183

F. Hydraulic profile of outlets 195

G. Reservoir and its hydraulic structures 197

H. Information on reservoir sedimentation 199

I. Infiltration and percolation 203

J. Results of the measurements in a paddy terraces block 205

K. Graphs of scenario simulations with RIBASIM 225

L. Summary 235

M. Samenvatting 241

N. About the author 247

Acknowledgements

A PhD study is like a study of life itself. I started to look for a PhD scholarship by downloading the guideline of the PhD booklet at the Nuffic website in 2004. I had a dream that after being married and giving birth to two babies, I would pursue my study at a higher level. Before being married I had one request to my ex boyfriend that is now my husband. I did not want to have a big house, jewellery, or anything else material but just one thing, his permission to continue my study. I met him when I was a lecturer at Udayana University, Bali, Indonesia. The reason was simple, while I am a teacher I should have more knowledge to share with my students. As well when in the bachelor study I had a dream to study in the centre of water engineering science, the Netherlands, where is the best non-gravity drainage system in the world.

In the PhD rules of the Netherlands Fellowship Programme (NFP), it is described that to become a PhD student at UNESCO-IHE, you should be graduated or have a master degree from UNESCO-IHE, or other institution in the Netherlands. Then I applied for a master study scholarship two times (2003 and 2004). Unfortunately, I did not pass it. Then in 2007, I got a NFP fellowship for a short course at UNESCO-IHE.

Furthermore, in 2009, I was awarded a scholarship from the Indonesian Government, and I had registered already in the University of Wollongong, New South Wales, Australia. At the same time, my father was sick, and he asked me to accompany him for 21 days in the hospital, because he should have a surgery in his spinal column. Because of this I cancelled my starting time to study in Australia. Afterwards I had the plan that I would start the study in 2010. Unfortunately, at this year, I cancelled again my planning, because my parents went to Mecca. I relied on my parents to accompany my children when I am studying abroad. My reason was natural, as a mother and a wife I should manage the household first, even though in reality there can be unpredictable developments. I just tried to manage it best.

Based on cancellation of starting times of my study, I thought deeply that I should change the type of study, from full time to a sandwich study. Then I applied for an NFP fellowship in early 2010. At that time, I finished all the administration and finances of the Indonesian Government Scholarship, which I had received after the announcement in

2009. I tried hard, in the middle of 2010 the announcement of NFP came on. I was awarded it and then I started my study in February 2011. I am very thankful that I had the opportunity of the best moments in life.

During the study, I went through, with many struggles coming unpredictable, even when I tried to obey the requirements of my supervisor and the instructions of the PhD guideline. A very challenging process, I cannot explain them one by one here, but these efforts sometimes made my spirit decreasing. As well, the dilemma came on from my family that my mother was sick from March until May 2015 in Indonesia. At the same time, I had to manage and write my draft PhD thesis in Delft. Then, she passed away on 15 May 2015. It was a hard time for me after losing her. Conversely, I accepted all struggles as learning process of life and PhD life itself.

Thank you very much I dedicate to my promotor Prof. Dr. Bart Schultz for his patience and thoughtful supervision of me during the PhD study. I am thankful that Prof. Dr. I. Nyoman Norken and Dr. László Hayde are my co-promotors. As well, I am thankful to Ir. W.N.M. van der Krogt, Ass. Prof. Ioana Popescu, Ass. Prof. Andreja Jonoski, Mrs. Claire Taylor, Dr. Krishna Prasad and Dr. Suryadi for their time to discuss with me. Additionally, not forgetting, I am grateful to the UNESCO-IHE staff that gave me the uniqueness of familiarity. I am very thankful to Ms. Jolanda Boots, based on her professionalism to take care of my PhD administration.

A big gratitude is given to I Made Semada as Leader of Subak Agung Yeh Ho River Basin (*Pekaseh* Subak Agung Yeh Ho) for giving all his indigenous knowledge of how to maintain social-religious, traditional technology in water distribution and maintenance, and traditional agro-economic elements in Subak irrigation schemes. Also thank you to Ni Kade Puspitasari, female staff of Public Works at the Telaga Tunjung Reservoir office, who supported me to collect the daily data in the paddy terraces block and the data on agricultural production at the sample blocks.

Moreover, I appreciate the officers at *BWS* Bali-Penida, Ministry of Public Works; Department of Public Works Bali Province; farmers of Subak Agung Yeh Ho, especially farmers of Subak Gede Caguh and Subak Gede Meliling; Indonesia Agency for Agricultural Research and Development (IAARD), Ministry of Agriculture; Department of Agriculture and Horticulture Tabanan; Soil Mechanic Laboratory of Bali State

Polytechnic, Geological Agency of the Indonesian Ministry for Energy and Mineral Resources to provide data for my research.

For the best, I am very blessed with the support of my dear husband, Ristono and our lovely son Narayana Radya Aydin (14 years old) and daughter Larasati Ridha Alisa (12 years old), without their loves and tolerants, I could not have gone through this process successfully. As well I am very thankful to my parents, first my father Suharto, who always encouraged me to pursue my study at the higher level. When I was studying for the bachelor degree, he advised me that as a woman, even when you become a wife or a mother you should have the existence of yourself. Subsequently you can transfer it to your children and your neighbourhood. Second, in memory of my deceased mother Roekminingsih, who was a teacher from 1961 till 1975, even if she decided to become a house wife, she always teached me how to become an independent person.

Finally I am grateful to my brothers Kukuh Wahyu Utomo and Winahyu Hadi Utomo, mother in law Ibu Sarmi, sisters in law and brother in law, all of my family, my UNESCO-IHE friends, especially Yenesew Mengiste Yihun, Mario E. Castro Gama, Wied Winaktoe Wiwoho, Bosman Batubara, Mona Delos Reyes, Reem F. Digna, Girma Yimer, Shahnoor Hasan, Jakia Akter, Claudia Quintiliani, Adey Nigatu, and Nguyen Thao. Also my roommates Fatma Balany, Hesti Nur Paramita, Junira Ardiana, Ernestasia Siahaan and Elizabeth Valentin, my best friends Widayati, Mei Lilia Triana, Tri Sulistyowati, Siti Nur Asmah, my colleagues, especially G.A.P. Candra Dharmayanti, Anissa Maria Hidayati, I Nyoman Sunarta and I Nyoman Udayana at Udayana University, and all my friends in Indonesia who supported me.

Delft, 2017

Mawiti Infantri Yekti

1 Introduction

Indonesia is one of the countries blessed with a large potential of water resources. Kardono (2005) stated that the potency of Indonesia's surface water is higher than 2,000 billion m^3/year, where Papua is in the first place with 1,400 billion m^3/year, followed by Kalimantan with 557 billion m^3/year, then Java with 118 billion m^3/year. The surface water scatters in 5,886 rivers, 33 million ha lakes, reservoirs and lowlands in 470 river basins, of which 64 are in a critical condition. The critical condition is due to several factors, namely pollution, change of land use, deforestation and agricultural activities within the river basins.

The seasons change every six months. The dry season (June to September) is influenced by the Australian continental air masses, while the wet season (December to March) is the result of the Asian and Pacific Ocean air masses. The air contains vapour that precipitates and produces rain in the country almost the whole year through.

During the twentieth century, rainfall reduced with 2 to 3% in Indonesia (Boer and Faqih, 2004). Almost all of this reduction occurred during the months from December to February. The precipitation patterns have also changed. There has been a decline in annual rainfall in the southern regions, an increase in precipitation in the northern regions and the seasonality of precipitation (wet and dry seasons) has changed. In addition, the wet season rainfall in the southern regions has increased, while the dry season rainfall in the northern regions has decreased (Case et al., 2008).

The changes in the wet and the dry season also affect the water availability, with decreasing rainfall during critical times of the year that may result in drought risk and uncertainty in water availability. Consequently, uncertain production of agricultural products, economic instability and more undernourished people, hindering progress against poverty and food insecurity (Wang et al., 2006). Other consequence may be that the water supplies to irrigation systems are affected, which is a part of water resources management, particularly in terms of sustainability of water supplies. That is also happening in one of the river basins in Bali called Yeh Ho River Basin.

The basin characteristics of Yeh Ho River in southern Bali show an elongated shape with the main river on the right side. Yeh Ho River has three reaches, which include upstream called the first time/part (*ngulu*), midstream called the second time/part (*maongin*) and downstream called the third time/part (*ngasep*). In the upstream, Yeh Ho River has its source, a spring called Gembrong. Since late 1990, the accounted discharge of some diversion weirs shows a reduction of discharge in the river. As a result, the distribution of water to the Subak irrigation systems has been disturbed (Regional River Office of Bali-Penida, 2006). This may have been caused by the fact that since 1987 the Bali Province Government, under the management of the Local Water Supply Utility (*Perusahaan Daerah Air Minum (PDAM)*) as regional-owned corporation (*Badan Usaha Milik Daerah (BUMD)*), has utilized the spring water for domestic purposes beyond its share of 65%. In 2001, in response to the claim of the Subak farmers in the upstream schemes, Tabanan Regency Government decided to restore the 35% allocation of Gembrong Spring for them under responsibility of the river basin organization, called Subak Agung Yeh Ho. However, this is not really followed in practice.

Associated with the shortages of land and water irrigation systems will experience the challenge in water resources management, especially in the distribution of water (MacRae and Arthawiguna, 2011). In fact, Indonesia faces challenges of water resources management until the present time. One of the challenges is application of the synergy concept top down and bottom-up. Indonesia has not been able to accommodate the public as part of a potential performer in problem solving, project-oriented use of technology to overcome the problems in water resources management, and insufficient involvement of the community in the finalization of planning and design criteria for hydraulic structures (Jayadi and Darmanto, 2011).

One type of irrigation systems that has existed for a thousand years are the Subak irrigation systems in Bali, which are farmer based irrigation systems with independent institutions. Historically irrigation, including Subak irrigation systems, has developed in Southeast Asia in four distinct but overlapping phases: community irrigation, river diversion dams, large storage dams and tube wells with pumps (Barker and Molle, 2004).

Subak irrigation schemes - related to which is the term of Irrigation Planning Criteria (*Kriteria Perencanaan (KP)*), called Irrigation Area (*Daerah Irigasi (DI)*) - are an example of water resources management, distribution and supply of irrigation water in a perfect vision on the social welfare in the river basin. The decision-making process of

Subak irrigation takes into account political, economic, social and cultural (religious) elements. Multifunctional ecosystems are implemented in a sustainable way of agriculture in the Subak irrigation schemes, particularly in the technology of these schemes.

The challenges, as mentioned above, are also experienced in the Subak irrigation schemes in Yeh Ho River Basin, where the Telaga Tunjung Dam started operation in 2006. Previously, the Subak irrigation systems in this river basin were supplied with water by diversion weirs at several points along the main river. The reason to build the dam was insufficient development of the water resources that caused a new problem among the farmers. In the past, the amount of water in Yeh Ho River was mostly utilized for the Subak irrigation and it was taken for drinking water as well in the upstream. The Subak irrigation systems are located both upstream and downstream of the reservoir. Each Subak irrigation system is operated and maintained by a group of farmers, who since the construction of the dam have problems to distribute the irrigation water. The present practice of reservoir operation based on rule curves, which are predominantly guided by direct economic benefits, while ignoring ecosystem requirements, needs to be reviewed (World Meteorological Organization (WMO), 2008). It may be noted that sometimes, with some minor operational and/or structural modifications, existing reservoirs can adapt the managed flows. However, it is recognized that maintaining (in the case of new reservoirs), or re-establishing (in the case of existing reservoirs) the natural river condition by the managed flow may not always be possible and may require a conflict management mechanism to determine the appropriate approach. Therefore, it is essential to develop a decision-making framework capable of handling conflicting demands and to apply an integrated approach.

Reservoir operation studies can be approached in different ways, such as by mathematical models and setting purposes of reservoir operation. Studies based on mathematical models were for example *Deriving reservoir operation rules via fuzzy regression and ANFIS* (Mousavi *et al.*, 2003); *Development of fuzzy reservoir operation policies using genetic algorithms* (Zahraie and Hosseini, 2007), *Harmonics elimination in multi level inverter using linear fuzzy regression* (Othman and Mekhaizim, 2010). These studies focused on the application of mathematic equations of simulation and optimisation in reservoir operation.

In the context of setting the purpose of reservoir operation within a river basin, especially for irrigation water, a decision support system for reservoir operation can be applied to be able to rotationally distribute irrigation water among sub-irrigation blocks. Then, by using this system, the operators can control the reservoir in the emergency state to possibly save water until coming rainy days (Kim *et al.*, 2003). An operational method for a regulating reservoir to effectively utilize excess water was examined for a canal system. The simulation could estimate quantitatively the effect of the regulating reservoir to make excess water effective (Nishimura *et al.*, 2005).

The central concern of various aspects of reservoir operation as well as to show possibilities how flows can be managed successfully to minimize their adverse impacts and optimise the benefits from ecosystem and socio-economic activities, implies that reservoir operation needs to focus at (World Meteorological Organization (WMO), 2008):

- understanding changes in flow and sediment regimes by reservoirs;
- identifying the issues that need to be quantified for deciding on the managed flows;
- introducing options to tackle these issues by modification of reservoir operation;
- planning the managed flows;
- providing a framework for environment sensitive reservoir operations.

Reservoirs need to provide for flow release to meet their specific purposes as well as the needs of the downstream ecosystem and livelihood objectives identified through scientific and participatory processes (International Union for Conservation of Nature and Natural Resources (IUCN), 2003). These flows are also referred to as 'environmental flows', which concern a certain minimum quantity of water released downstream of a reservoir. They have to serve certain environmental objectives, which need to be understood and defined. Several approaches are available for assessing the environmental needs of river systems downstream of a reservoir. According to Olivares (2008) the engineering objective solution to environmental impacts of hydropower operations on downstream aquatic ecosystems were studied using a revenue-driven approach by a linear programming model optimisation for operation of a reservoir after the Bay Hydropower Complex. The model was formulated and solved for parametrically varying levels, or environmental constraints. It was attempted that the managed flows would maintain the natural flow and sediment regimes variability as close as possible. Therefore, the main objective of managed flows is to compromise in the allocation of water between releases and retaining sufficient water in the reservoir to support economic activities, e.g. water supply, hydropower and flood moderation, for which the reservoirs were originality built, or the purposes for which new reservoirs are proposed (Figure 1.1).

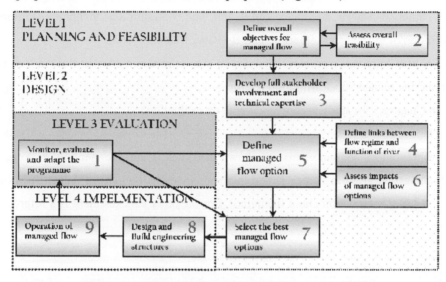

Figure 1.1. Managed flow approach (Acreman, 2000)

In connection with the construction of the Telaga Tunjung Reservoir, this research aims at optimising the relation between reservoir operation and the operation of the Subak irrigation systems in the river basin in such a way that these systems can be maintained and even be further developed for their agriculture productivity. The thesis is organised in eight chapters.

Chapter 1 provides the introduction.

Chapter 2 presents background of the research related to Subak irrigation and the objectives for this research.

Chapter 3 provides information about the development of Subak irrigation schemes related to learning from experiences over a thousand years of participatory irrigation system management in ancient Subak irrigation schemes of Bali.

Chapter 4 presents information on Subak irrigation in the south of Bali with the challenges to manage the flow for these irrigation systems. This is followed by a discharge analysis for a system approach to river basin development with Subak irrigation schemes as a culture heritage.

Chapter 5 provides information on the interaction between irrigation and drainage of Subak irrigation schemes in light of the farmer's perspective.

Chapter 6 shows a model simulation and optimisation technique to enable optimisation of the managed flows in Yeh Ho River Basin.

Chapter 7 presents a scenario analysis on the optimisation of river flow in Yeh How River and the related Subak irrigation schemes.

In Chapter 8 the study is concluded with an evaluation.

Abbreviations and acronyms are shown in Annex A and the symbols in Annex B. The Summary is given in Annex L and the 'Samenvatting' (Dutch) in Annex M. Finally Annex N gives information about the author. For the other annexes there is a reference in the text.

2   Background and objectives

2.1 Profile of the region

Bali is one of the 17,504 Indonesian islands and one of the 34 provinces. It is surrounded by 25 named islands and 60 unnamed islands as shown in Figure 2.1 (Department of Internal Affairs, 2004). Administratively, the Province of Bali is divided into 8 districts and 57 sub districts with 1 city and 710 villages, which have 3.9 million people. Bali is part of the Lesser Sunda Islands, it has 153 km length, 112 km width and an area of 563,000 ha (0.29% of the Republic of Indonesia). Astronomically, Bali is located between latitude 8°25'23" South and longitude 115°14'55" East.

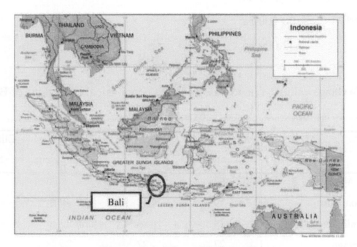

Figure 2.1. Map of Indonesia

(http://www.lib.utexas.edu/maps/middle_east_and_asia/indonesia_rel_2002.pdf)

Bali has a tropical climate, like the other parts of Indonesia. It has a dry season from April to October and a rainy season from November to March. The average annual rainfall varies between 1,000 and 3,000 mm (Regional River Office of Bali-Penida, 2011). Bali has a minimum temperature between 23.0 and 24.5 °C, the average temperature is between 26.4 and 27.3 °C and the maximum temperature between 29.5 and 31.8 °C. The month of July has the coolest temperature. Wind velocity of Bali fluctuates over the year between 9.3 and 13 km/day, the relative humidity is approximately between 69 and 93% (Tables 2.1 and 2.2).

The topography of Bali has a slope in northern and southern direction with the mountains in the middle (Figure 2.2). The mountains and hills cover 85% of the area. The slopes 0 - 2%, 2 - 15% and 15 - 20% are located in the districts of Badung, Tabanan, Gianyar, Buleleng and in the coastal areas, while areas with slopes above 40% extend from the middle to the South (Regional River Office of Bali-Penida, 2011). There are two active volcanoes namely Batur Mountain (1,717 m+MSL (mean sea level)) and Agung Mountain (3,142 m+MSL) and 22 inactive volcanoes. If the total area of 563,000 ha of Bali is distinguished by slopes the areas are as follows (Wapedia, 2011):

- 0 – 2%   :   123,000 ha;
- 2 – 15%   :   118,000 ha;
- 15 – 40% :   190,000 ha;
- > 40%   :   132,000 ha.

Table 2.1. Average values of meteorological and geophysical conditions of Bali in 2008 by station

	Meteorology Ngurah Rai	Geophysics Sanglah	Geophysics Karangasem	Climatology Negara
1. Temperature (°C)				
• maximum	30	32	30	30
• minimum	24	24	23	23
• average	27	27	26	26
2. Relative humidity (%)				
• maximum	93	89	89	93
• minimum	71	68	69	71
• average	83	79	79	82
3. Air pressure (mm Hg)	757	758	745	758
4. Wind velocity (km/day)	11.1	9.3	13	13
5. Rainfall (mm/year)	1,790	1,890	2,210	1,720
6. Sunshine (%)	75	72	65	67

Source: Central Bureau of Statistics, 2009

Table 2.2. Meteorological and geophysical conditions of Bali in 2008 by Regency

Regency/ Municipality	Temperature (^oC)	Relative humidity (%)	Rainfall (mm/year)	Wind velocity (km/day)
Jembrana	26	82	1,720	13
Tabanan	19	90	4,110	111
Badung	27	83	1,790	111
Gianyar	27	80	2,120	111
Klungkung	28	89	2,190	111
Bangli	28	83	2,480	111
Karangasem	26	79	2,210	13
Buleleng	27	81	1,340	13
Denpasar	27	79	1,890	93

Source: Central Bureau of Statistics, 2009

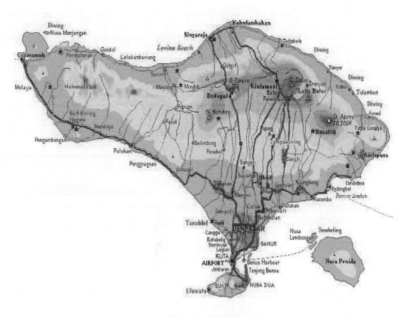

Figure 2.2. Map of Bali

(http://www.baliguide.com/bali_map.html)

The 15 - 40% land slope area has soil fertility characteristics that depend on the source rocks and weathering rates, while the 2 - 15% land slope area has the characteristics of sediment transporting alluvial rivers (Romenah, 2010). The potential lands for agriculture are spread in Bali and are mostly found in the mountainous and hilly areas. Due to the good soil fertility over the years the crop yields in Bali have been quite stable, with the harvested area in the range of 142,000 - 160,000 ha (Central Bureau of Statistics, 2011). This is evidenced by the stability of agricultural products, especially rice, with yields in the range 787,000 - 879,000 tons/year, which is approximately 5 tons/ha of unhusked rice or 2.85 tons/ha of rice (Figure 2.3). This is achieved by the farmers of the Subak irrigation schemes.

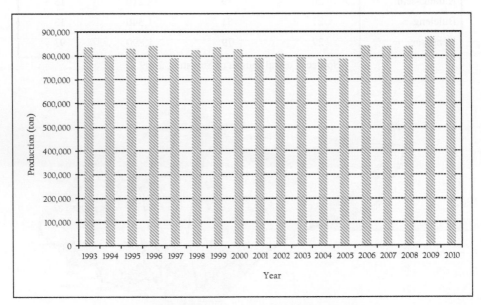

Figure 2.3. Production of rice in Bali Province
(Central Bureau of Statistics, 2011)

Indonesia is an agricultural country, with Bali that is known as the rice granary. Accordingly, the agricultural production became a mainstay in this island, in addition to the tourism industry. In particular industry is increasing, as a result of the development of world economy towards a free market and more competition. Nevertheless, the number of people in Bali who work in agriculture, estate crops, forestry and fishery still occupies with 726,000 people a top position. Approximately 790,000 people work in several other sectors such as mining and quarrying, manufacture, electricity and water supply, construction, transportation, storage, communication, finance, insurance, real estate and public services. About 482,000 people work in trade, restaurants, hotels and the tourism industry (Central Bureau of Statistics Bali Province, 2009).

In the mid of the twentieth century, as consequence of ongoing population growth and land conversion, Balinese farmers were having difficulties in meeting the ever-growing demand for rice. The Government constructed large dams to increase water supplies based on the available hectares of paddy fields. However, due to this also the problem of insufficient water in the dry season developed.

2.2 Water resources

2.2.1   World water resources

The world's water exists naturally in different forms and locations: in the air, on the surface, below the ground and in the oceans. Only 2.5% of the Earth's water is fresh. Nearly 68.7% of this water is frozen in glaciers and ice sheets. The 0.4% of world's surface and atmospheric water is divided in 67.4% in freshwater lakes, 8.5% in lowlands, 12.2% in soil moisture, 1.6% in rivers, 9.5% in the atmosphere and 0.8% in plants and animals (Figure 2.4). The reduction in available water resources depends on human activity and natural forces. Even though public awareness of the need to better manage and protect water has grown over the last decades, economic criteria and political considerations still tend to drive water policy at all levels. Science and best practices are rarely given adequate consideration (UNESCO, 2006). Pressures on water resources will continue to increase primarily because of urbanization, population growth, increase in living standards, growing competition for water and pollution.

However, a large volume of freshwater exists 'in storage'. It is therefore important to evaluate the renewable annual water flows, taking into account where and how they move through the hydrological cycle (Figure 2.5). The scheme of the hydrological cycle illustrates how elements can be grouped as part of a conceptual model that has emerged from the discipline of eco-hydrology, which stresses the relationships and pathways shared among hydrological and ecological systems (Zalewski *et al.*, 1997).

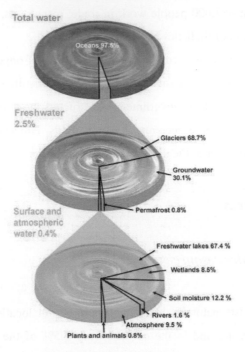

Figure 2.4. Global distribution of the world's water (UNESCO, 2006)

Figure 2.5. Schematic view of the hydrologic cycle components (UNESCO, 2006)

2.2.2   Water resources in Indonesia

Although Indonesia is blessed with abundant water resources, it still faces various challenges with respect to these resources. Therefore, it is important to update information on water resources periodically. AQUASTAT presents the water availability information of Indonesia as shown in Table 2.3 (Food and Agriculture Organization of the United Nations (FAO), 2005). Total actual renewable water resources (TARWR) is the sum of renewable water resources and incoming flow originating from outside the country. It is a measure of the maximum theoretical amount of water annually available for the country. Derived from the data of FAO, Indonesia only uses 3% of the TARWR of which the highest percentage of 91% for withdrawal of water for the agricultural sector, as follows:

- total freshwater withdrawal (km^3/yr)   :   82.8;
- per capita withdrawal (m^3/p/yr)   :   372;
- agricultural use (%)   :   91;
- domestic use (%)   :   8;
- industrial use (%)   :   1;
- agricultural use (m^3/p/yr)   :   339;
- domestic use (m^3/p/yr)   :   30;
- industrial use (m^3/p/yr)   :   3.

Withdrawal typically refers to water taken from a water source for use. It does not refer to water 'consumed' in that use (FAO, 2006). The agricultural sector includes water for irrigation and livestock. The domestic sector typically includes household and municipal uses as well as commercial and governmental water use. The industrial sector includes water used for power plant cooling and industrial production.

According to Naylor (2007), agriculture is central to human survival and is probably the human enterprise most vulnerable to changes in climate. This is particularly true in countries such as Indonesia, with large populations of rural and urban poor. Understanding the current and future effects of changes in climate on Indonesian rice agriculture will be crucial for improving the welfare of the country's poor.

Table 2.3. Water resources in Indonesia

Population: 223 million
Precipitation rate: 2,700 mm/year
Total actual renewable water resources (TARWR)
Volume in 2005: 2,840 km³/year (2,840 billion m³/year)
Per capita in 2000: 13,400 m³/year
Per capita in 2005: 12,800 m³/year
Breakdown of total actual renewable water resources:
• surface water: 98%
• groundwater: 16%
• overlap is water shared by both the surface water and groundwater systems: 14%
Incoming waters: 0%
Outgoing waters: 0%
Total use of total actual renewable water resources: 3%

Source: FAO-AQUASTAT, 2005

2.2.3   Water management in Indonesia

The main challenge of water management in Indonesia will be oriented to minimize the conflict of interest on water management among the regions within a river basin, law enforcement and implementation of regulation, increasing community participation and awareness, capacity building in dealing with environmental issues, and improvement of the performance of water allocation planning and conservation activities. Hence, the development of water management policies requires evidence-based information that deals with complex, contextual and multi-aspect issues (Alexander *et al.*, 2010). In addition, sector oriented management, top-down approach, and illegal/illegitimate management activities are indications of crisis water resources management in Indonesia (Jayadi and Darmanto, 2011). Therefore, a solution is needed to integrate the natural system (both quantitative and qualitative) with the societal system of the resource users (Loebis, 2003).

The most recent law on water resources, which conforms to the decentralization legislation, was passed by Parliament on 19 February 2004. It is known as Law No. 7/2004 on Water Resources. This law is based on Constitution 1945, Art. 33 Para (3) and Peoples Consultative Council (*Majelis Permusyawaratan Rakyat (MPR)*) Decree No. IV/MPR/1999 about the Outlines of State Policy (*Garis-garis Besar Haluan Negara (GBHN)*). This law formulates the mission of water resources management with water resources utilization, water resources conservation, water-related disaster control, empowerment and improvement of Government, public and private participation, water resources data viability and accessibility and advancement of information systems (Directorate of Water Resources and Irrigation, 2005). This law was an improvement of the Law No. 11/1974, which still supported centralization of decision-making in water resources development.

In 2010, the Government formulated the future challenges in its National Policy on Water Resources, including the main role of water resources conservation, based on the Millennium Development Goals (MDG) and the Johannesburg Summit of 2002. This was the basis of the implementation of technology in water resources, and the coordination and synchronization at national, provincial, district, and river basin level to establish a reliable institutional water resources management (Jayadi and Darmanto, 2011). The answers to challenges in water resources management focused on:

- to overcome the problem of the water crisis due to the uneven distribution of population and economic activity from one island to another;
- to reduce the potency of low social performance, economic and environmental loss consequences due to the low performance of water resources management.

Because of some inconsistencies in the law of 2004 on 18 February 2015 the law was repealed. This brings for the moment the legal position back to the law of 1974. The Minister of Public Works and Public Housing, Basuki Hadimuljono, expects that Law Number 11 of 1974 will be adapted where required and then accepted. The old law is simpler than the new law, which provides sufficient scope for the Government to adjust design regulations with respect to water management to the up to date conditions (Bimantara, 2015).

2.2.4   Water resources in Bali

Bali is blessed with a large water potential, especially in the southern region. Based on the Regulation of the Minister of Public Works number: 13/PRT/M/2006, July 17th, Bali Province has the Bali-Penida River Region that is divided in 20 Sub-regions. The 20 Sub-regions have 165 river basins in which there are 49 seasonal rivers (renewed by Government Regulation (*Peraturan Pemerintah*) number 12 of 2012, code 03.03.A3, about the establishment of Bali-Penida River Region that covers the area of the Province of Bali, and divides it in 391 river basins with an area of in total 5,617 km^2). In the middle of Bali there are four lakes, namely Lake Beratan, Lake Buyan, Lake Tamblingan and Lake Batur, 1,273 scattered springs and eight groundwater basins.

Due to global climate change, the rainfall during the rainy and dry seasons affects the hydrological cycle. Therefore, it becomes quite difficult to predict it (Case *et al.*, 2008). Bali as one of the regions with agricultural stable products is also experiencing such changes. The data need to be updated by the Regional River Office of Bali-Penida of the Department of Public Works. The annually potential water of Bali as described in the river basins map (Figure 2.6) consists of 4,126 million m^3 of river flow, which is from surface runoff and interflow, 781 million m^3 of springs and 252 million m^3 of groundwater. Then the total is 5,160 million m^3.

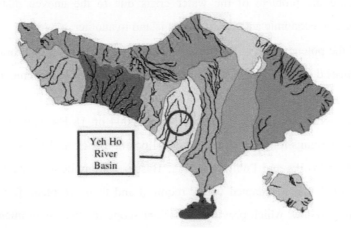

Yeh Ho River Basin

Figure 2.6. River basins map of Bali

(http://galerigis.com/Peta-Tematik/Peta-Das/Peta-Das-Bali)

Four large reservoirs exist in the Bali-Penida River Region. The Palasari Reservoir started operation in 1989, the Grokgak Reservoir in 1998, the Telaga Tunjung Reservoir in 2006 and the Benel Reservoir in 2009 (Figure 2.7). The main purpose of all reservoirs is to fulfil the needs of Subak irrigation water.

The area of lakes and reservoirs has not changed over the last years and is equal to 3,588 ha (Raka, 2009).

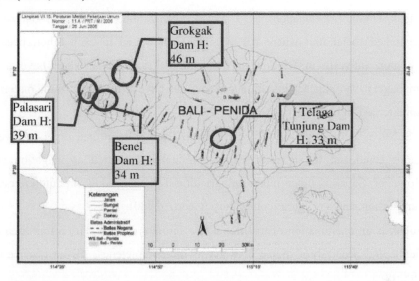

Figure 2.7. Dams in the Province of Bali (Regional River Office of Bali-Penida, 2011)

2.3 Definition of Subak irrigation schemes

Bali has a tradition of managing irrigation water by Subak irrigation schemes since the 9th Century (Norken *et al.*, 2010). If we mention the word Subak, generally Bali's people will interpret it in one of the following descriptions (Sutawan, 2008):

- a region of rice fields with an area and certain limits;
- the rice farmers who are gathered in one organization that is engaged in the management of irrigation water;
- physical system or irrigation system itself with canals (*telabah*), traditional diversion weirs, secondary/tertiary boxes and other facilities.

Furthermore, there are definitions of Subak given by several researchers and Subak observers as follows:

- Subak is of indigenous people in Bali who are socio-religious agrarians. It was historically established since ancient times and continues to grow as a ruling organization of land in the field of water management and others for the rice fields from a water source within a region (*PERDA No. 02 / PD/DPRD/1972*);

- Subak Union (*Persubakan*) as a social organization, called *Seka Subak*, is an organized social unit where the members feel bound to each other, because of common interests in relation to irrigation of rice fields. They have a leader (manager) who can act inside or outside and have a treasurer, both material and immaterial (Sutha, 1978);

- a Subak is defined as all terraces irrigated from a traditional diversion weir and major canal and the term Subak is commonly translated as irrigation society. However, Subak is much more: an agricultural planning unit, an autonomous legal corporation, and a religious community (Geertz, 1980);

- Subak is an organization of traditional rice farmers in Bali, with a unit area of rice fields, and generally a source of water as the completeness of the essence (Kaler, 1985);

- a Subak irrigation system is in addition to a physical system also a social system. The physical system is defined as the physical environment closely linked to irrigation water as the source and irrigation facilities in the form of diversions or dams, canals, secondary or tertiary boxes, etc., while the social system is a social organization that manages the physical system (Sutawan, 1985);

- Subak is defined as an organization of water users for fields of its members to obtain water from the same source, having one or more of the temples near secondary diversion structures (*Ulun Bedugul*) and having full autonomy well into the care of the interests of their own households, or in the sense of the word freely entered into direct relationships with outside parties independently (Sutawan *et al.*, 1986);

- Subak is more precisely called as a socio-technical-religious organization system rather than called socio-agricultural-religious organization system (Arief, 1999);

- a Subak irrigation system is besides an appropriate technological system, also a cultural system. The phenomena indicates that basically a Subak irrigation system is a technological system that has been developed as a part of a cultural society (Pusposutardjo, 2000);

- a Subak system is a custom law community with socio-technical-religious characteristics. It consists of a group of farmers that manage irrigation water at their paddy fields (Windia *et al.*, 2000);

- Subak may be defined as a socio-religious agriculture and irrigation institution dealing primarily with water management for the production of annual crops, particularly rice, based on the *Tri Hita Karana (THK)* philosophy (Sutawan, 2002);

- Subaks are recognized in social anthropology as irrigators' associations that combine ritual and resource management (Jha and Schoenfelder, 2011);

- as a system Subak is widely known as a 'traditional' irrigation management institution for rice cultivation on the Indonesian island of Bali (Roth, 2011). Subak and irrigated rice agriculture became well adapted to, and embedded in the characteristic Balinese landscape of rugged mountains and steep valleys deeply incised by fast-flowing rivers.

Considering the various above definitions of Subak, it may be observed that none of them covers the full context of a Subak irrigation scheme. Therefore underneath definition has been developed. This definition reads (Yekti *et al.*, 2012):

'A Subak irrigation scheme, primarily in Bali, Indonesia is an irrigation system of which the construction, operation and maintenance is based on agreed principles of technology, management of agriculture and religious community'.

2.4 Subak schemes

2.4.1   Paddy terraces

Most of the Subak cultivation areas are paddy terraces. Paddy terraces are elements in irrigation systems that offer a beautiful view for tourists who visit Bali. However, not only Bali has traditional irrigation systems (Subak). Especially in Asia almost all countries have rice fields irrigated by traditional irrigation institutions similar to Subak, including India, Korea, Nepal, Oman, the Philippines and Thailand. Definitely, their names are different according to local terms, for example *Khul* in India and Pakistan (*Karez* in Baluchistan), *Falaj* in Oman, *Zanjeras* in Philippines and *Muang-fai* in Thailand (Coward and Levine, 1987). According to Von Droste *et al.* (1995), the paddy terrace landscape is a cultural landscape and unique characteristic of Asia's countries. Therefore, these landscapes would have to be protected. Meanwhile Indonesia has traditional irrigation systems other than Subak, such as *Panriahan-pamokkahan* in North Sumatra and *Panitia Siring* in South Sumatra. The cultural landscapes in every corner of the world are the result of interactions between humans and nature. Diverse people living in different environments and with different cultures have developed strategies to survive in their landscapes, creating numerous forms of such landscapes in their efforts to sustain their communities (Luchman *et al.*, 2009).

In 2003 the Subak landscapes were nominated for the World Heritage List of UNESCO (Fowler, 2003). This was also strengthened by Bridgewater (2003), who stated that the major factor in the identification and maintenance of cultural landscapes is the understanding of the world-views that have shaped them. The Subak landscapes were officially added to the list on June 29, 2012, as a manifestation of the *Tri Hita Karana* philosophy and reflecting their significance for sustainable development.

The abundance of agricultural fields is possibly the most obvious characteristic of Bali. The large areas of fertile land and abundant water resources have permitted paddy cultivation for a long time, and rice cultivation has become one of the main economic activities. This has developed over the centuries in the specific socio-cultural, agro-ecological and political administrative environment of this mountainous island. As a consequence, the Subak irrigated rice agriculture became well adapted to, and embedded in, the characteristic Balinese landscape of rugged mountains and steep valleys deeply incised by fast-flowing rivers. In many parts of the island, such as sloping upland areas, the land is fragile and sensitive to disturbance because of agricultural activities. Rice terraces have been created as a strategy to permit the use of hilly and sloping environments (Whitten *et al.*, 1996).

The key issue for the future is what policy settings are needed to ensure their survival in the face of environmental homogenization, as part of the general process of globalization.

2.4.2   *Subak irrigation system*

The Balinese have build diversion weirs (*empelan*) in the rivers or canals to irrigate their paddy fields (Figure 2.8), without worrying about the flow of water through the fields, because the excess water was discharged into drains, and was then available for irrigation of other fields. Since the 9th Century, Subak Associations have been in charge of the management of the Subak irrigation schemes by using these diversion weirs. Through these weirs, irrigation water was supplied to the respective schemes and within the schemes distributed to each Subak Association member by using water distribution units (WDU/*tektek/kecoran*). This condition allows farmers (individual) and Subak Associations to have access to each other's irrigation water (Windia *et al.*, 2006). Therefore, Subak irrigation is a farmer based irrigation system with an independent association (self-governing irrigation association). The independent institution has certain rules and procedures of work, which are related to design, technology implementation, water management and operation and maintenance of the overall canal systems and structures (Pribadi and Wena, 2007).

Bali's Subak irrigation is known as a gathering of farmer organizations with determination and high spirits to work together (*gotong royong*) in the efforts to obtain water for producing food crops, especially rice and secondary crops. As an institution of traditional farmer irrigation, rice cultivation activities have existed in Bali since 882. The word *huma*, that means cultivating rice field in Indonesian, was mentioned in epigraph Sukawana (Purwita, 1993; Sutawan, 2008). Whether the *huma* in question is cultivating irrigated paddy fields, is also evidenced by the word of *undagi pangarung* in Bebetin epigraph written in 896, which means tunnel maker (*arungan/aungan*) (Sutawan, 2008).

Furthermore, the Pandak Badung epigraph was found in 1071 that has the word *kasuwakan,* meaning *kasubakan* or Subak (Purwita, 1993; Sutawan, 2008).

Figure 2.8. Paddy terraces of Subak (Yekti *et al.*, 2013)

The format of Subak irrigation system management is remarkably homogenous throughout Bali. Variations are a result of unit size or regional naming systems (Birkelbach, 1971). Poffenberger and Zurbuchen (1979) stated that in the history of pre-mechanized agriculture few societies have ever achieved the high levels of productivity characterized by wet-rice farming in Bali. With traditional technology, the Balinese farmers could produce twice as much rice on their land as their neighbours, the Javanese farmers, whose techniques were by no means unsophisticated. How have the Balinese farmers done it? It appears that four factors are central to their success as rice farmers. These include the fertility of the volcanic soil, a complex technology and corresponding knowledge of wet-rice cultivation, which allows the Balinese farmers to make maximal use of environmental systems and resources. As an organizational system, Subak irrigation has the capability of coordinating workers and resources, and genetic strains of rice selected during a thousand years for their disease resistance, productivity and beauty.

However, Subak Associations still have some problems with the condition of certain water control structures (diversions) that are generally made of simple wood constructions. Consequently, some improvements can be made with respect to the quality of Subak irrigation, especially related to the hydraulic performance of the systems (Suanda and Suryadi, 2010).

2.5 Subak cultivation area

The area of paddy wet fields under Subak cultivation in Bali reached 85,700 ha in 2000. When viewed from the type of irrigation, the irrigated rice fields reached 98.8%. It means that most of the rice areas are wet paddy fields. However, in 2008, the paddy field area decreased to 81,400 ha (Table 2.4) (Central Bureau of Statistics, 2009). This decrease was caused by land conversion to residential and industrial areas, or other types of land use.

2.6 Previous studies on water management of Subak irrigation schemes

Previous studies on Subak irrigation can be described based on three main subjects; these are related to perspective, organization, and history and technology. The perspective of the Subak irrigation schemes based on the *Tri Hita Karana (THK)* philosophy as an equivalent technology in irrigated agriculture was presented by Windia *et al.* (2002). This study discussed the form of Subak irrigation schemes as an appropriate technology, implemented in the form of thinking-pattern, social system, and the development of the system. The final goals of the schemes are to achieve harmony and togetherness in irrigation management. The second study discussed the perspective issues of agrarian change in South-Central Bali (Lorenzen, 2010). According to him, although rice farming continues, for many households it has become a side business. The flexible nature of rice farming in terms of labour input and available casual off-farm work allows farming households to allocate their available labour to a variety of on-farm and off-farm income generating activities. On the other hand, a significant part of the younger generation is unwilling to work in the 'mud' and there is little appreciation of the many benefits the Subak provides not only to the farming community, but also to the wider community.

Furthermore, on the technological elements it was explained by Gany (2004) that in Indonesia, there remains a lot of phenomena of the ancient heritage of participatory irrigated agriculture practices adhered to Subak irrigation, that need to be uncovered in terms of scientific explanation. An ethnographic study undertaken by Pribadi and Wena (2007) aimed at providing information about the irrigation structures and technologies as scientific description of specific human cultures in Bali.

Table 2.4. Area and type of agricultural land use in Bali in 2008 (ha)

1		Agriculture land	355,800
1.1.		Paddy wet field	81,400
	a.	Full technical irrigation	160
	b.	Semi technical irrigation	71,000
	c.	Simple irrigation by Public Works Department	5,100
	d.	Traditional irrigation	4,600
	e.	Non irrigation	490
	f.	Tides rise and fall based	-
	g.	Lowland	-
	h.	Polder and others	-
1.2.		Non paddy wet field	274,400
	a.	Field (*Tegal/Kebun*)	136,700
	b.	Field (*Ladang*)	-
	c.	Estate crops	121,700
	d.	Wood land	10,400
	e.	Sea fish pond	680
	f.	Fish fond	290
	g.	Grassland	-
	h.	Temporary not used	230
	i.	Others	4,400
2		Non agriculture land	207,200
	a.	House, building, and land surroundings	44,700
	b.	State forest	123,400
	c.	Swamp	100
	d.	Others	39,000
		Total:	563,000

Source: Central Bureau of Statistics, 2009

Another study on these elements had the objective to achieve better farming, which needs some improvements in the scientific quality of Subak irrigation, especially related to the hydraulic performance (Suanda *et al.*, 2010). A fourth study was conducted based on an inverse technique to find out if the ability of Subak irrigation can be transformed. Then by the fuzzy set theory the dominance or ranks of the elements of Subak irrigation can be determined, which are also a consideration in the transformation process (Windia *et al.*, 2010).

A study of the organization and history of Subak organizations and their efforts to improve as irrigation institution has been done by Norken *et al.* (2010). Another study presented how water shortages are ascribed to the dominance of the tourism industry, private companies selling bottled drinking water and regional water delivery services in Tukad Ayung River, all of which farmers hold responsible for crop failure in dry years (Straub, 2011). Straub focused on the perspective of Subak members during water scarcity caused by lack of coordination between privatized and previously centralized water resource management based on economic priorities for the tourism sector and urban regions, and water use for agriculture.

A model of the complex adaptive system of Balinese water temples by Lansing and Kremer (1993) was developed to solve complex coordination problems in allocating water and to control pests. The coordination problem can be solved by assuming the local rules how individual communities make their decisions. Claims that have been made by Lansing for explanatory power of this model are unwarranted. Therefore, there can be doubt about the relation of the modeland the real-world events (cf., Elster, 1986; Smith, 1995; Vayda, 1995a). Janssen (2007) tested this model and found that the robustness of their insights was determined by the ability of agents to self-organization, which was sensitive to pest dynamics and assumptions of agent decision-making. According to his opinion, if water availability is the key problem, it is wise to coordinate with others not to use water at the same moment and the coordination among Subak farmers will require more comprehensive behavioural rules than imitating the best neighbour.

2.7 Problem description

The study area is managed by Subak Agung Yeh Ho. Yeh Ho River has from upstream to downstream 11 weirs, being: the new diversion weirs Aya and Penebel, the traditional diversion weirs: Benana, Riang and Sigaran, and the new diversion weirs: Jegu and Caguh in the upstream, the new diversion weir Meliling I in the midstream, and the new Gadungan, Sungsang I, and Sungsang II weirs in the downstream. The cropping patterns have been determined and agreed upon by Subak Agung Yeh Ho, which have been used and are believed as fair under the traditional system of water distribution.

In the period 2003 - 2006, Meliling II Weir in the midstream has been replaced by Telaga Tunjung Dam, which enabled storage and changed the function of the hydraulic structure from a weir to a dam. Telaga Tunjung Reservoir with a surface area of 16.5 ha is located in Village Timpag, Kerambitan District, Tabanan Regency, Bali Province. The distance is 40 km northwest from Denpasar, the capital of the province. The dam is located downstream of the confluence of Yeh Ho River and Yeh Mawa River.

In this case, the purpose of the weir was to elevate the water level in order to enable gravity flow to the Subak irrigation systems, and the purpose of the dam is besides to elevate the water level, to store water as well. Furthermore, the river flow downstream changed quite significantly due to a change in water sharing since the reservoir became in operation (Figures 2.9 and 2.10).

The following points contribute to the role of reservoir operation as managed by an independent association in sustainable irrigation water supplies to Subak irrigation systems managed by farmers, who can also develop their productivities in agricultural income:

- shortage of water has developed;
- increased land conversion causes reduction in agricultural area and affects the river basin;
- Subak farmers need to adapt their practices to the changed distribution of water from diversion by a weir to storage and distribution from a reservoir.

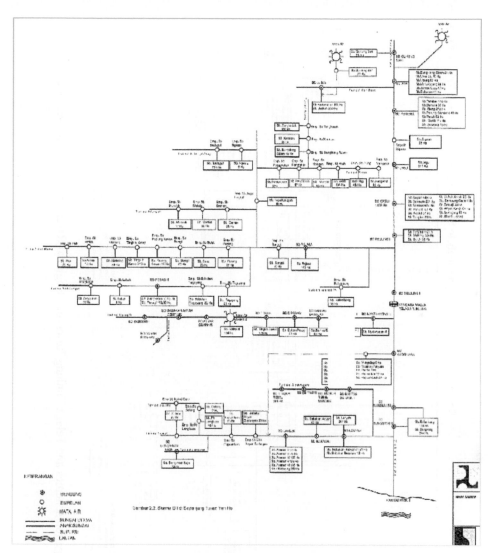

Figure 2.9. Subak irrigation schemes in Yeh Ho River Basin

(Department of Public Works, 2005)

Reservoir operation for water supply to Subak irrigation schemes in Yeh Ho River Basin

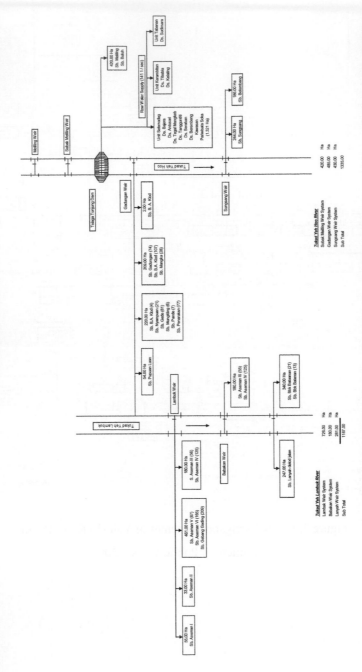

Figure 2.10. Subak irrigation schemes and Telaga Tunjung Reservoir (Regional River Office of Bali-Penida, 2006)

2.8 Objectives

2.8.1   Research questions

Since the reservoir was built in the midstream of Yeh Ho River due to the change in river flow downstream quite significant changes in the process of water sharing occurred among the Subak irrigation schemes. Moreover, due to utilization of the Gembrong Spring for domestic needs since 1987, there is a serious conflict on the water distribution among the farmers within the river basin. This resulted in the following research questions:

- how can a Subak irrigation scheme result in sustainable productivity and even be further developed under limited water resources?
- can new operation and maintenance technologies be applied in Subak irrigation systems in order to improve agricultural production?
- which reservoir operation rule can best manage the flow within the river basin, and how would it have to be applied by the river basin agency?
- how do farmers perceive the changes in the operation pattern of the water distribution from the reservoir, while previously the water was distributed by a weir?
- what is the subsistence of the reservoir in terms of water conservation?
- how can the reservoir be operated to allocate the irrigation water in an optimum way?
- how can the related Subak irrigation schemes best be operated in relation to the reservoir operation?

2.8.2   Hypotheses

The hypothesis of this study is that it is possible to manage water resources in a sustainable way particularly by managed flows of Subak irrigation systems within a river basin, which is influenced by the potential water resources, water needs and characteristics of a reservoir and its position in the river basin.

It is expected that managed flows within a river basin can be analysed in the form of modelling and scenario analysis, where the factors described above are formulated in a system approach that may result in a reservoir operation that is able to improve the agriculture productivities in the Subak irrigation schemes.

This hypothesis is related to the objective of managed flow under Integrated Flood Management (IFM) and Integrated Water Resources Management (IWRM) that would have to be: 'To imitate the natural flow conditions, as far as possible, in order to minimize the adverse impacts on the livelihoods of people through eco-services' (World Meteorological Organization, 2008).

The main purpose of building a reservoir for irrigation in the humid tropics is to store water in the rainy season and to supply it in the dry season. The first characteristic is the reservoir capacity. Mass curve analysis shows the fluctuating curve in the Rippl diagram method (1883) as describing the mass supply, while the constant line is describing the mass demand. The required storage capacity of a reservoir is the vertical mass a+b (Figure 2.11).

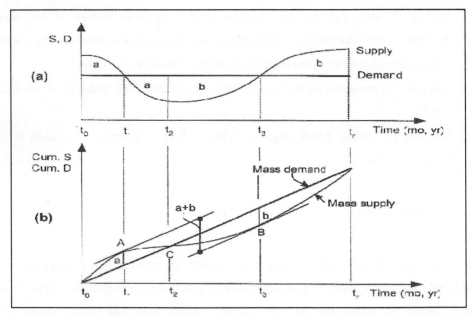

Figure 2.11. Mass curve analysis (Rippl diagram method, 1883; Akintug, 2010)

The discharge data of the midstream of Yeh Ho River have been considered. The line of cumulative mass is more or less constant (Figure 2.12). Due to the construction of the Telaga Tunjung Reservoir with an effective capacity of 1 million m^3 a modified operation and maintenance of the involved Subak irrigation systems, by changing it from traditional operation of the diversion weir to reservoir operation (dam) is needed.

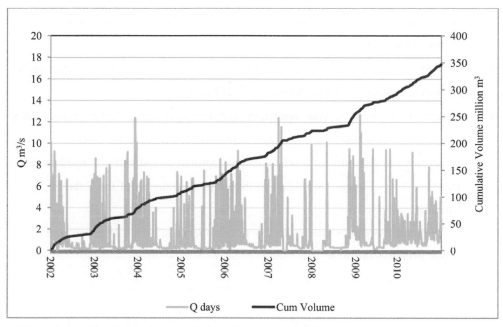

Figure 2.12. Discharge characteristics midstream of Yeh Ho River in the South of Bali (Regional River Office of Bali-Penida, 2011)

2.8.3   Research objectives

The main objective of this study is to develop an optimal reservoir and related Subak irrigation systems operation, capable to support agricultural productivity at upstream, midstream and downstream level.

From a scientific perspective, as shown in Figure 2.11, it is of importance whether the river requires storage of water or not, which depends on the river flow characteristics. Therefore, study was needed how the reservoir can best be operated. The research needed to be conducted by considering the river basin as a whole system, which in this study is the river with several Subak irrigation systems. Therefore, detailed objectives of study that have been addressed are to:

- identify the type of river basin and to evaluate the land use;
- identify the contribution of the existing hydraulic structures in the main river system in supplying water to the Subak irrigation schemes;
- identify and analyze the reliability of discharge in the main river system and the inflow to the reservoir;
- analyze and determine the optimal outflow from the reservoir based on the needs of the Subak irrigation schemes within the river basin;
- simulation of the reservoir operation and the operation of the related Subak irrigation systems based on the needs of the Subak irrigation schemes at upstream, midstream and downstream level to achieve optimal productivity of agriculture in a sequence time of operation related to the cropping patterns of the Subak irrigation schemes;
- to formulate recommendations on the future operation of the reservoir and related Subak irrigation schemes.

3 Development of Subak irrigation schemes: learning from experiences of ancient Subak schemes for participatory irrigation system management in Bali

3.1 Introduction

Subak irrigation systems have been well known since the 9th Century. These systems are managed by a Subak Association based on the *Tri Hita Karana (THK)* philosophy, as a faith of Balinese-Hindus based on harmony between people and nature, harmony between people and people, and harmony between human beings and God. This philosophy underlies every activity of Subak farmers. For managing the Subak systems, Subak Associations and farmers pursue the Subak regulation called *Awig-awig* Subak as the togetherness consensus that was originally announced by the King and nowadays by the Head of the Regency. As associations for irrigation system management, Subak Associations have been already naturally adapted to Participatory Irrigation Management (PIM). This chapter is based on a literature study to portray centuries of experience with ancient Subak irrigation system management, in which PIM was represented by three linked elements: PIM in irrigation system operation and maintenance; PIM with respect to socio-culture and economics of agriculture; PIM in light of a religious community. While several of these systems are now under stress, the results of this literature study may hopefully contribute to sustainable PIM for the operation and maintenance of Subak irrigation schemes in Bali during the forthcoming decades.

A special aspect of the paddy terraces landscapes is the water management in the Subak irrigation schemes as shown in Figure 3.1 (Yekti *et al.*, 2012). The successful management by applying the Subak irrigation practice distinguishes Balinese terrace management from other terrace management practices. Hence, the harmonious nature of the Balinese cultural landscape based on the *THK* philosophy presents an excellent model for sustainable development (Luchman *et al.*, 2009).

The *THK* logo in Figure 3.2 implicitly contains a message for us to manage water resources wisely in order to maintain their sustainability. The *THK* philosophy of irrigation system management has three subsystems: (i) material subsystem (including technology); (ii) social subsystem (including economy); (iii) cultural subsystem (way of thinking, norms and principles). All subsystems have a balance with the environment as shown in Figure 3.3.

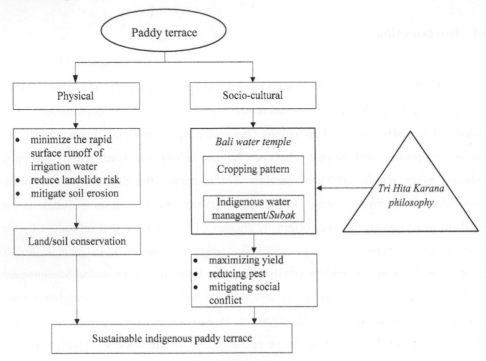

Figure 3.1. The characteristics of the paddy terraces as a cultural landscape in Bali bound to the Balinese *THK* philosophy (Luchman *et al.*, 2009)

Figure 3.2. *Tri Hita Karana (THK)* (Yekti *et al.*, 2014)

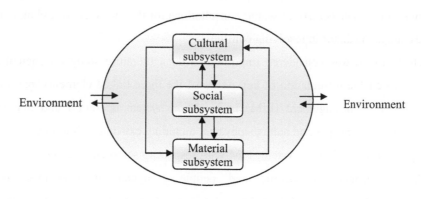

Figure 3.3. Links between subsystems in the socio-cultural community for irrigation system management (after Arif, 1999)

As an association of landowners, that got the auspices from the Government, the Subak Association has its regulation (*Awig-awig* Subak) (Arga, 2011). This regulation was and still is applied to prosper the community based on the *THK* philosophy. For Subak farmers, the *Awig-awig* Subak is a tool for management of the irrigation system.

The developing times can influence human behaviour and thinking patterns. This also concerns the Subak farmers. Lorenzen *et al.* (2005) pointed out that all levels within the Subak systems are autonomous and interdependent at the same time and showed that even the formal regulations are still existent. Farmers at the lower level fine-tune the system by relying on informal arrangements.

Present and future challenges faced by Balinese Subak farmers are: i) increasing competition in marketing of agricultural products due to trade liberalization; ii) declining interest of the rural youth to become farmer; iii) financial burden due to operation and maintenance in the tertiary, quarternary, etc. level; iv) decline of irrigated land areas due to conversion to other uses, such as tourism and industry; v) degradation of the environment, limited availability of water resources and competition with other water users; vi) the irrigation systems and hydraulic works are partially damaged; vii) ownership of a farmer's land is 0.5 ha or less, making it difficult to achieve a decent life; viii) lack of coordination and supervision from the Government since the lack of clarity by the Regent (Great *Sedahan*), and the Head of the District (*Sedahan*) who were instrumental in coordinating the management of water, consultations for the managers of Subak irrigation schemes and as mediator in resolving conflicts.

Therefore, it was considered important to do a literature study on ancient Subak irrigation, to enable formulation of key issues of the three linked elements: participatory irrigation system management (PIM) in irrigation system operation and maintenance (harmony between people and nature - physical including existing technology: the terraces landscape and the hydraulic system, PIM with respect to socio-culture and economics of agriculture (harmony of people and people - agreement/compromise between Government/stakeholders and farmers within socio-culture and economic elements in agriculture, including regulations and organizations, and PIM in light of a religious community (harmony of human beings and God - faith/belief in God in the ceremonies and hierarchy of Subak temples as core lake ecosystem/water is belonging together and a blessing).

3.2 Method and discussion

In the initial period, people together in a community opened farmland and needed the irrigation water to be distributed, arranged and managed from sources to their farmland. Then Subak Associations were formed. In 1072, the first writing on Subak has been recorded (Goris, 1954; Norken *et al.*, 2007). However, as agricultural wetlands (rice fields) they have already existed before the 11th Century (Purwita, 1986; Norken *et al.*, 2007). Moreover, it cannot be denied that the people have been participating independently in communities related to sharing water by using socio-technical elements as Subak irrigation schemes since long time ago.

Irrigation schemes are basically based on social and technical elements (Huppert and Walker, 1989; Pusposutardjo, 1997b). A statement that irrigation schemes have a socio-technical nature was supported by the Government Regulation (*Peraturan Pemerintah*/PP) 77/2001. While Subak irrigation schemes are based on the *THK* philosophy, it also means that it are systems of socio-technical nature, in which the technology is fused with the socio-cultural elements of local communities. The character of the Subak technology has been identified by Poespowardojo (1993), as the technology that has evolved into the culture of the people. The form of *THK* in the management of irrigation water in Subak irrigation schemes is shown in Table 3.1.

Table 3.1. Socio-technical nature of the *Tri Hita Karana* philosophy in Subak irrigation scheme management (after Windia *et al.*, 2005)

Contents of *Tri Hita Karana*	Implementation of *Tri Hita Karana*
1. Cultural subsystem 1.1 Religious elements (*Gatra parhyangan*)	• water is considered highly valued and respected, and is created by the One Almighty God • the temple is a place of worship of the One Almighty God, and regarded as part of the control mechanism for managing irrigation water • regular religious ceremonies
1.2 Human relation elements (*Gatra pawongan*)	• irrigation water management based on the concept of harmony and togetherness
1.3 Natural relation elements (*Gatra palemahan*)	• special land is provided for holy buildings (temples) at locations where it is considered to be important • the remaining land near distribution boxes is used for holy buildings (temples), so conflicts over land can be avoided
2. Social subsystem 2.1 Religious elements (*Gatra parhyangan*)	• there is a regulation called *Awig-awig* • accountability of irrigation water management • rights of water and land are respected • there is a system of additional water for downstream called *pelampias* in irrigation water management
2.2 Human relation elements (*Gatra pawongan*)	• there is a flexible structure of the Subak Association • there is mutual cooperation and payment of contribution to succeed the Subak activities • there are routine meetings of the Subak Association
2.3 Natural relation elements (*Gatra palemahan*)	• the Subak's members permit the building of holy buildings (temples) at remaining land near distribution boxes

Table 3.1. *continued*

3. Material subsystem 3.1 Religious elements (*Gatra parhyangan*)	• water flows continuously through the distribution boxes. It is managed by using this system that is overseen by the One Almighty God, also there has been presence of temples near the distribution boxes • a concept of water distribution unit with continuous flow called *tektek* in any of the distribution boxes of the Subak systems is considered to distribute the irrigation water fairly and proportionally • mutual cooperation between the members of the Subak Association and the ordinary members, so the Subak programs can be completed by harmony and togetherness
3.2 Human relation elements (*Gatra pawongan*)	• cooperation between the members of the Subak Association and the ordinary members, so the Subak programs can be done by harmony and togetherness • coordination between Subak leaders with other stakeholders in the area, such stakeholders are customary village, village agency, government agencies and others, in order to achieve that the Subak programs are properly implemented
3.3 Natural relation elements (*Gatra palemahan*)	• topography of Subak schemes is typically sloping area • every paddy block of one farmer has one inlet and one outlet • boundaries of Subak schemes are naturally clear • presence of hydraulic structures and irrigation systems are suitable with the needs of the local farmers • use of local materials in support of the irrigation systems

Scientifically it still needs to be explored how the *THK* philosophy has been and has to be implemented related to the Subak irrigation schemes. The *THK* philosophy in Subak irrigation is a faith of Balinese-Hindus to maintain all elements in agriculture. However, Roth and Sedana (2015) stated that the scientific and policy concepts in irrigated agriculture and the Subak domain as an ideology are not simply based on local wisdom, tradition or culture but that it requires critical scientific scrutiny as part of wider processes of socio-political change.

The etymologically meaning of the *THK* philosophy is the love of truth or the love of wisdom instead of an ideology, which is a systematic assemblage of ideas, believes and truths to provide direction of communities in various aspects of life, as politics, law, defence and security, social-culture and religious activities. Starting from the *THK* philosophy the goal of scientific engagement can further be developed without limits.

3.2.1   *PIM in irrigation system operation and maintenance*

As stated before, for centuries, the Balinese have built diversions in rivers or canals to irrigate their paddy terraces fields without worrying about the flow of water through the fields. This condition allowed individual farmers and Subak Associations to have access to each other's irrigation water. Prior to the Majapahit era, these activities were done independently. During the Majapahit Kingdom (1343 - 1500) there was Government influence by the head of the Subak regency (*Asedahan*), who served to coordinate water arrangements among the Subak irrigation schemes and coordinated collection/tribute/ incentive of land taxes (*sewenih/tigasana*). After the Majapahit Kingdom this was done by the head of the district (*Sedahan*) until the Dutch Colonial period with regency government officers (*Sedahan Agung*), the Government intervened related to coordination of operation and maintenance (Norken *et al.*, 2010). In the Dutch Colonial period, a start was made with the building of permanent hydraulic structures (*empelan*), like weirs and other supporting structures, such as primary and secondary canals and discharge measurement structures, such as Cipoletti and Thomson weirs.

After independence, starting in 1986, the standard of irrigation systems was completely described in the Irrigation Planning Standards (*Standard Perencanaan Irigasi*) in seven series of books with Planning Criteria (KP) by the Department of Public Works. Moreover, the infrastructure and facilities for irrigation have been discussed related to the system and the operation and maintenance of it within the context of a river basin in line with the most recent laws and regulations, such as Government Regulation (*PP*) 20 / 2006 on irrigation, and Act number 7/2004 on water resources, that was cancelled in 2015. Therefore, since then in fact activities have to be based on the laws and regulations preceeding to this law.

Subak irrigation schemes consist of simple irrigation structures. The general technology of the hydraulic structures of the schemes consists of (Yekti *et al.*, 2013): i) a diversion/intake (*buka or bungas*), or weir (*empelan*); ii) primary canals (*telabah gede*), secondary canals (*telabah pemaron*), tertiary canals (*telabah cerik*), quarternary canals (*telabah pengalapan*) to distribute the water to several owners of paddy fields. If one quarternary canal belongs to 5 owners, it is called *telabah panca*, and if one canal belongs to 10 owners, it is called *telabah penasan*. The canal that is used to distribute the water evenly within one small field is called *talikunda*; iii) primary boxes (*tembuku aya*), secondary boxes (*tembuku pemaron*), tertiary boxes (*tembuku cerik*), and boxes to distribute water among several owners of paddy fields (*tembuku penyahcah*). If one box belongs to 5 owners, it is called *tembuku panca*, and if one box belongs to 10 owners, it is called *tembuku penasan*. The box to distribute water for one owner is called *tembuku pengalapan*; iv) water distribution unit (*tektek/kecoran)* (Figure 3.4); v) to break through hills there may be tunnels (*aungan*); vi) the structure at the end of a tunnel is called *kibul*, drain (*pengutangan*), and receiving water body (*pangkung*) (Figures 3.5 and 3.6).

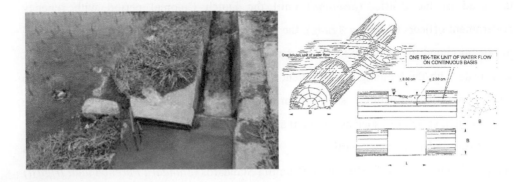

Figure 3.4. Typical water distribution unit (*tektek or kecoran*)

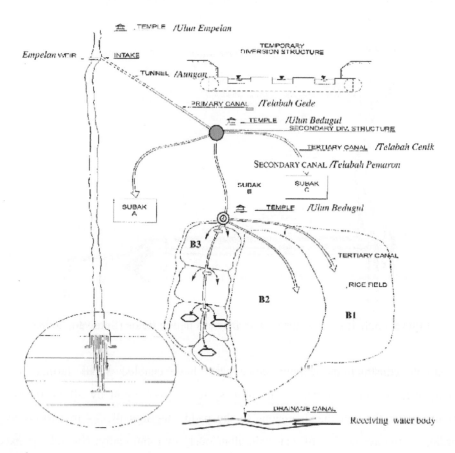

Legend:

⬤   = primary box (*tembuku aya*)

◎   = secondary box (*tembuku pemaron*)

▭   = tertiary box (*tembuku cerik*)

◇   = box to distribute water for many owners of paddy fields (*tembuku penyahcah*)

B1, B2, and B3   = water users associations (*tempek*) in Subak B

Drain (*pengutangan*)

Receiving water body (*pangkung*)

Figure 3.5. Subak irrigation systems with hydraulic structures from the source to the paddy fields (after Gany, 2004)

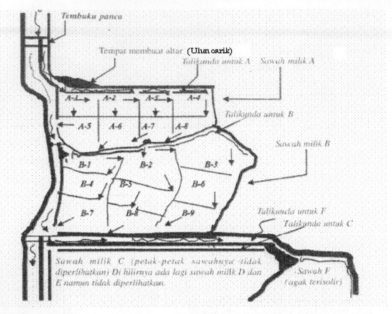

Figure 3.6. Irrigation water flow in farmer's paddy fields (Sutawan, 2008)

Simple constructions without sluice gates have enabled Subak farmers to: (i) regulate water based on a water allocation system agreed upon by all members; (ii) distribute water through the diversions to the paddy terraces blocks in various ways, according to the availability of water simultaneously or rotationally; (iii) release excess water to drains (*pengutangan*). Therefore, members of a Subak Association have three activities namely (Yekti *et al.*, 2014):

(i)   *allocation and distribution of irrigation water.* Irrigation water allocation is an activity that entitles the use of available water to each member behind a water distribution unit (WDU, *tektek/kecoran*) with the following dimensions: 5 - 8 cm of width and 1 - 2 cm of height. A section of rice fields with an area of about 0.3 - 0.4 ha is supplied from the main/secondary/tertiary/quarternary canal (Figure 3.7) under: (i) a continuous flow system, or (ii) an agreed scheduling system (Sutawan, 2008). The scheduling system may consist of:

*   scheduling based on rotation of the cropping pattern (*megilir*);

*   scheduling based on the starting time of land preparation in the paddy cultivation season, which is called *nyorog/nugel bumbung*;

* scheduling based on the season: wet season called *masa* and dry season *gadon*;

* scheduling based on the period of water use right in a year (*tebak taun*) or the period to plant rapid growing rice (*tebak cicih*);

(ii) *the transfer mechanism among the members of the Subak Association behind a single weir, within a Subak irrigation system, and between the Subak irrigation schemes* (Figure 3.7). The transfer procedure of irrigation water is based on mutual dealing among water users and is coordinated by the leader of the Subak irrigation subsystem called *Pekaseh* and for a total Subak irrigation system called *Subak Gede*. All Subak irrigation schemes within a river basin are called Subak Agung. Here the coordination is by the leaders of the large Subak irrigation schemes called *Pekaseh Subak Gede* and the leader of Subak irrigation schemes in a river basin, called *Pekaseh Subak Agung*. In the past the auspices of Government rested with the King.

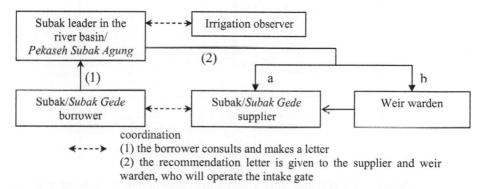

Figure 3.7. Transfer procedure of irrigation water along the Yeh Ho River

(iii) *procedure for operation and maintenance of the irrigation systems.* Implementation of Article 18 of the Ministry of Public Works and Public Housing Regulation No. 04/PRT/M/2015 on Criteria and Designation of Irrigation Status is based on irrigation areas for decision-making in which the Government manages the irrigation systems at primary and secondary level. Thus, the construction of new canals, weirs, reservoirs or other structures and the financing of operation and maintenance at the primary and secondary level are under government responsibility.

Although the Government has the responsibility for operation and maintenance of the primary and secondary canals, it sometimes uses the Subak farmers to maintain the main system. On the other hand, if there is damage in the lower level system, the Subak Association can request for a grant from the Government. Even though the grant is not launched immediately, the Subak farmers still can solve their technical problems with their indigenous knowledge and individual financial resources (Figure 3.8).

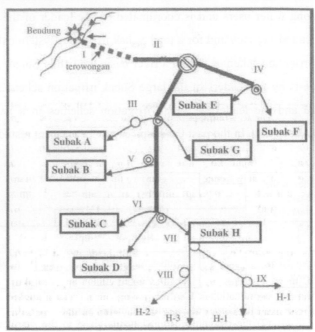

Figure 3.8. Segregation of duties among Subak members in minor financing of physical improvement and irrigation system maintenance

The Subak Associations have a mechanism for minor financing of physical improvement and maintenance at lower level as shown in Figure 3.8. The operation and maintenance of water distribution units is controlled by occasionally using intermittent irrigation. However, the water in the primary, secondary, tertiary and quarternary canals still flows to the lower level schemes. Subak farmers have the duties of operation and maintenance at the lower level (Figure 3.8), starting from the tertiary canals and tertiary boxes (II), quarternary boxes (III, IV, V, and VI), and the boxes for 5 or 10 landowners (VII, VIII, and IX).

The indigenous regulation for institutional management and water sharing (*Awig-awig* Subak) has survived to cope with the traditional conditions for more than a thousand years. The *Awig-awig* Subak describes all aspects related to water distribution and law for Subak members. As long as the water sharing is agreed upon by the members of the Subak Association, although it may be scientifically not quite accurate, they will accept it as a fair and equitable water sharing.

Subak Associations obtain financing from two sources: i) external budget from government grants; ii) internal budget from fees and independent business. Each Subak Association has its own way to get the budget to operate and maintain the physical systems at the tertiary and/or lower level. Overall, the funding process is sufficiently flexible that farmers may direct government funds as they see fit, and they invariably claim to channel funds towards the uses that they prioritise. Even with the more heavy-handed government agendas during the Suharto era (1967 - 1998), farmers often imposed their own agendas using strategies ranging from overt resistance to more subtle 'everyday forms of resistance' (Scott, 1985; Pedersen and Dharmiasih 2015).

3.2.2   PIM with respect to socio-culture and economics of agriculture

As socio-cultural and economic agricultural organizations Subak Associations had already a complete institutional system. In addition, Subak Associations had the areas and the water sources of their own. Thus, it is reasonable that the Subak Associations are regarded as autonomous organizations that run their irrigation systems accordingly based on the Subak regulation. The contents of the regulation are proposed by the Subak farmers, who are the actors in the field.

The authorities did not interfere in the construction, operation and maintenance of the irrigation schemes. The Subak members conducted this independently, without external financial support. In the Kingdom era, the King merely gave permission to his people to crush the forest and to open new fields, as well as, to raise the river flow for irrigation by building weirs *(empelan)*. The farmers were required to pay by using the tribute of crops. The Subak autonomy was widespread, as revealed by the fact that each

Subak Association had its own regulation, which was defined by democratic consensus that could be different from others, including their organizational assembly. The assembly of a Subak Association is shown in the Figures 3.9 and 3.10.

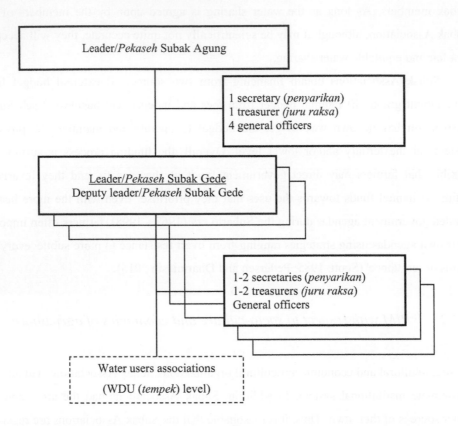

Figure 3.9. Association structure of Subak Agung Yeh Ho (after Sutawan, 2008)

Also in 1908 during the Dutch colonial period, very high authority was given to the Regency Government officers in their territory to coordinate the District Government officers in order to increase the contribution for the Dutch Colonial Government (Norken *et al.*, 2010). The aim was to increase revenues from farmers' tax on land, and the selection of Subak leaders was done by the Regency Government officers or District Government officers. The financial accountability of the Subak Associations was controlled as well by these officials (Graders, 1939). During the Dutch colonial period the role of the head of Subak regency (*Asedahan*) was transferred to the supervisor of the Subak irrigation schemes (*Sedahan/Sedahan Yeh/Penglurah*) at the district level, and the head of Subak regency became the supervisor at regency level (*Sedahan Agung*).

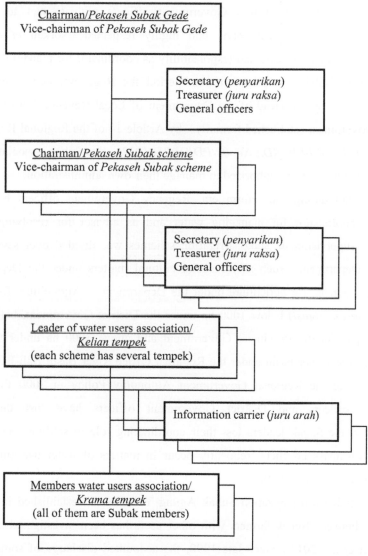

Figure 3.10. Association structure of *Subak Gede* in Tabanan Regency

(after Sutawan, 2008)

After the proclamation of independence in 1945, one of the most significant roles of the Regency Government officers and District Government officers was in the field of water distribution management among the water sources and among the Subak irrigation schemes. Normally the Subak members were obedient to the decisions of the Regency Government officers and District Government officers with respect to the water management. They were charismatic and highly respected (Norken *et al.*, 2010). Regency's Government officers had responsibility as coordinator for managing water and as advisor for resolving conflicts. In this period the Regency's Government officer function was held by the Head of the Department of Local Revenue (DISPENDA). In 1972, the Government established formally - by Article 14 of the Regional Regulation of Bali Province (*Pasal 14 PERDA No. 02/PD/DPRD/1972*) - that Subak Associations as the traditional institutions were authorized to manage their own administration.

After 1998 during the reform era, Regency Government officers had no role anymore as coordinator for managing water and as advisor for resolving conflicts. Because the supervision of Subak irrigation schemes was divided over several of the responsible Departments, such as for social cultural matters under the Department of Culture and Tourism, for agriculture under the Department of Agriculture, for irrigation schemes (*Daerah Irigasi*) below 1000 ha under the District Government, between 1000 and 3000 ha under the Provincial Government and above 3000 ha under the Central Government, for a river basin under the Regional Office River Basin Bali-Penida. Since the launching of the Regional Government Autonomy Policy in 2000 the Regency Government officers and District Government officers have lost their power. Consequently, the Subak leaders lost their coordinating role in settling down disputes. Therefore, nowadays conflicts frequently occur in matters of water use among Subak members (Norken *et al.*, 2010).

The river basin organization Subak Agung Yeh Ho was established in 1991. The competition between Subak farmers with other users has been investigated in this river basin (Yekti *et al.*, 2012). Since late 1990, the accounted discharge of some diversion weirs showed a reduction of discharge in the river, as a result the distribution of water to the irrigation systems was disturbed (Regional River Office of Bali-Penida, 2006). This may have been caused by the fact that since 1987 the Bali Province Government, under the management of the Local Water Supply Utility (*Perusahaan Daerah Air Minum (PDAM)*) as regional-owned corporation (*Badan Usaha Milik Daerah (BUMD)*), has utilized the spring water for domestic purposes beyond its share of 65%. In 2001, in response to the claim of the Subak farmers in the upstream schemes, Tabanan Regency Government decided to restore the 35% allocation of Gembrong Spring for them under Subak Agung Yeh Ho. However, this is not really being followed in practice.

The claim of the Subak farmers became fragile, because the value of irrigation water is significantly lower than the prices of domestic water and of hydropower. Furthermore, it will have a significant impact on the price of unhusked rice that is pegged by the Government, compared to the price of consumption rice in the market. According to the Central Bureau of Statistics Bali Provinsi (2013), per October 2013, the selling rate of unhusked rice by the farmers was Rp 3,923 per kg, and Rp 4,013 per kg for rice millers, while the Cost of Goods Sold (*Harga Pokok Penjualan*) of rice set out by the Government was Rp 3,300 per kg, and the minimum price of consumption rice in the market was approximately Rp 9,000 per kg.[*]

Before 2015, the Governments of Bali Province and the regencies were charged to implement Law No. 7/2004 on Water Resources and had to form a Water Resources Council for the Subak irrigation schemes, namely an Irrigation Commission. The re-introduction of the Regency Government officer and District Government officer functions as supervisors from the Government authority was required. This was in accordance with the Public Works Ministry Rule (*Permen*) 31/PRT/M/2007 regarding the Irrigation Commission Guidelines (Norken *et al.*, 2010). During this period, to face the challenges of water resources development, the authorities in charge of the coordination of the Subak irrigation schemes within a river basin (*Subak Agung*) and of the individual Subak irrigation schemes (*Subak Gede*) were appointed as members of this commission. Accordingly, the Subak farmers had a voice to give their opinion related to everything that occurred in the field. Therefore, decisions that were announced by the Government were also based on the voice of the Subak farmers.

---

[*] Rp 10,000 = US$ 0.96 average price level for 2013

As said before the Water Resources Law was cancelled because it did not meet the six basic principles of water resources management restriction. One of these, the fifth principle related to the customary rights of indigenous people who depended on water resources, was also recognized in accordance with Article 18B (2) of the 1945 Constitution. The regulations about inauguration unity customary law community are still alive through the Regional Regulation (*Peraturan Daerah-Perda*). It does not have a constitutive, but it a declarative status.

With respect to this it is also important that water resources utilization for other countries was principally not allowed. The Government only allowed giving the permission to other countries if the water supply to their own people's needs had been sufficient. The needs referred to daily needs, sanitation system, agriculture, energy, industry, mining, transportation, forestry and biodiversity, sports, recreation and tourism, ecosystems, aesthetic and other needs. Based on all the foregoing considerations the rights on authorization over water resources by the State is the 'soul' or 'heart' of the Law No. 7/2004 on Water Resources as mandated by the 1945 Constitution' (Anjasari, 2015). However, this does not influence the Subak irrigation that is based on the *Awig-awig* Subak, indeed the Subak farmers tended to have a voice in the arrangement of the Irrigation Commission during the period 2007 - 2014.

Furthermore, the paddy terraces landscape is a famous destination of tourism in Bali. In future, Bali tourism ultimately depends on the preservation of their rich and harmonious culture and natural landscapes. While tourism can potentially damage the island's ecosystems, the indigenous *THK* philosophy promotes sustainable development in several ways. First, the *THK* philosophy provides rules and guidelines by which humans can live in the biosphere in a sustainable manner. Second, the Balinese system is able to facilitate indigenous knowledge preservation and its application in the community. Nowadays, indigenous techniques such as ethno-ecology are widely practiced to manage the island's resources, and significantly contribute to the sustainable development agenda (Martin, 1995; Dudley *et al.,* 2005; Mercer *et al.,* 2007; Luchman *et al.,* 2009).

To make the livelyhood of the Subak farmers more economic, it is important to undertake such practices and creating indigenous community support for the sustainability of resources, especially with respect to agricultural products, use of in Subak irrigation schemes cultivated products in the hotels, and other forms of cooperation. The increasing income for farmers related to the prices of agriculture products requires Government oversight together with the role of private sector fairly, based on the third element of the *THK* philosophy, harmony between people and people.

3.2.3   PIM in light of a religious community

The religious element is still kept by the Subak farmers in their activities. Generally, the ritual activities are conducted collectively and individually (Sutawan, 2008). The temple of the Subak religious community is the place where the collective ritual activities are conducted. It concerns that they appreciate God Almighty *(Sang Hyang Widhi Wasa)* for granting the sources of water and the locations to where the water is distributed. The routine of collective ritual activities that they do in these places expresses that they are always reminded by God to maintain and share the water fairly.

The Subak collective and individual ritual activities are very diverse. All ritual elements are related to agriculture activities. The collective ritual activities are conducted by Subak members within cross-districts and regencies, and related to sustainability of water sources such as: ritual for hope on success for the period of the paddy session *(ngusaba)*, ritual for hope on success for the full session *(pakelem)*, ritual at the beginning of diverting water from the river to the primary canal *(magpag toya)*, ritual for the hope of God's blessing of the farmers *(ngerestiti)*, and ritual for the hope to be prevented from pest attacks *(nangluk merana)*.

The individual ritual activities are conducted by the Subak farmers at the lowest level, related to the activities during the planting season such as: ritual just before cultivating the land *(ngendagin)*, ritual after ploughing the land *(ngerasakin)*, ritual starting to plant seeds *(pengwiwit)*, ritual for choosing the best day for planting *(nuansen)*, ritual for calling and hoping for cleaning of weeds *(ngulapin)*, ritual 12 days after planting *(ngeroras)*, ritual for spreading slurry and hoping for fertility *(mubuhin)*, ritual one month after planting and hoping that attacks of pest can be prevented *(neduh)*, ritual two month after planting *(nyungsung)*, ritual when rice is flowering *(nyiwa seraya)*, ritual just before paddies begin to bear fruit *(biukungkung)*, ritual just before harvesting and hoping for the best result *(mebanten manyi)*, ritual after storing paddies in a granary *(mantenin)*, and ritual for hoping that God will bless the farmer *(ngerestiti)*.

All the religious activities are conducted vertically from the top to the lowest level of the hierarchy of the Subak temples and depend on the kind and the time of the ritual to be carried out. The schedule of the ritual activities is in line with the cropping patterns within the river basin (*Subak Agung*) and the individual irrigation systems (*Subak Gede*). The ritual activities, contribute to unity of the Subak members, and are expected to reduce conflicts that may arise with their activities in agriculture. Subak farmers believe the philosophy that water is belonging together, the philosophy that water is a blessing, and the coordination with others will sustain the Subak ecosystem.

According to Surata (2003) in the four lakes of Bali - Lake Batur, Lake Beratan, Lake Buyan and Lake Tamblingan - Subak temples, called *Ulun Danu* Temple, became the core of the management of the ecosystem itself, while Bali is formed by the management of the lake ecosystems. The *Ulun Danu* Temple is located on the edge of Lake Batur, and of Lake Beratan for three lakes (Beratan, Buyan and Tamblingan). The *Ulun Danu* Temple for Lake Batur is shown as number 1 in Figure 3.11, and for Lake Beratan for three lakes by the numbers 1 and 2 in Figure 3.12.

According to Lansing (1987), there are four types of hierarchy of Subak temples or water temples (*Pura* Subak) in the Bali Regencies as shown in the Figures 3.11 and 3.12. The hierarchy of the Subak temples is connected to the Subak irrigation schemes (technology of the hydraulic structures of the schemes, see the first aspect of PIM). Near the weir (*empelan*) is the temple called *Ulun Suwi* as number 5 in Figure 3.11a, near the primary and secondary boxes/diversion structures are temples called *Ulun Empelan* as number 6 in Figure 3.11 and *Ulun Bedugul* as number 7 in Figure 3.11, and near the distribution units (*tektek*) is a temple called *Ulun Carik* (number 7 or 8 in Figures 3.11 and 3.12 of the hierarchy of Subak temples).

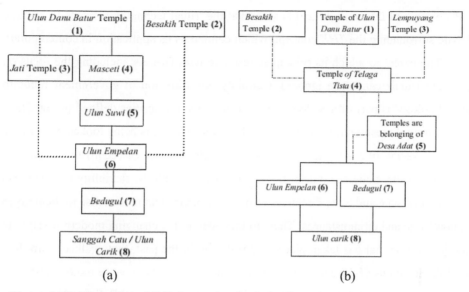

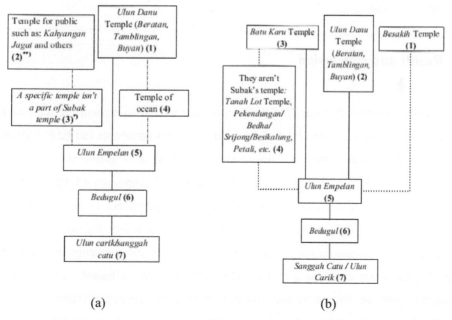

(a)   (b)

Figure 3.11. Hierarchy of Subak temples for Lake Batur in Bangli, Gianyar and Badung

Regencies (a) in Karangasem Regency (b) (after Sutawan, 2008)

(a)   (b)

Figure3.12. Hierarchy of Subak temples for the three lakes in Buleleng Regency (a) and in

Tabanan Regency (b) (after Sutawan, 2008)

Pedersen and Dharmiasih (2015) state that the 'ritual technology', needs to reframe the debate surrounding the *THK* concept, which condemns its application in contemporary agendas. The extent to which its principles resonate with farmers, including the focus on temples and rituals cannot be ignored. Certainly, the direction of government funds to 'ritual technology' is part of a State project and it is about priorities defined by the State, including cultural preservation and tourism. However, when it is being looked at from the ground up - 'seeing like a farmer' as shown in Stephan Lorenzen (2008) thesis title - it accords with farmer beliefs and priorities. Farmers work with the flexibility that they can have within the State and State directives to make decisions that make sense to them given their knowledge and circumstances. They do this even in the changing modern lifestyle in human civilization that has taken place. For instance, in the past, they walked from their house to the locations of the ceremonies, and in the modern era, they use motor cycles and vehicles. Yet, in some places and for specific ceremonies, some Balinese-Hindus still walk to the location of the ceremony, even though they have motor cycles and vehicles. They believe that the *THK* philosophy is capable to adjust in wise value to be relevant in their aspects of life.

3.3   Result and conclusion

This chapter illustrates that the *THK* philosophy as principle value and the regulation of the Subak Association (*Awig-awig* Subak) as togetherness consensus can be consistent by developing and changing the three linked elements: PIM in irrigation system operation and maintenance, PIM with respect to socio-culture and economics of agriculture, and PIM in light of a religious community. From the Kingdom era to the present, some behaviour and thinking patterns have changed. This can affect the socio-culture and economics of agriculture in the Subak irrigation schemes and the religious community. Moreover, there are concerns how the values of the *THK* philosophy can be applied consistently, although there have been dynamic changes in human civilization.

The lesson learned from the ancient Subak schemes is to have natural resources of topography, especially in the paddy terraces landscape, water resources and soils based on the principles of ecosystem management by using the *THK* philosophy. This philosophy also influences all the activities of participatory irrigation management related to those three linked elements in Bali.

Accordingly, Subak farmers irrigate their paddy terraces based on continuous flow related to the natural hilly topography, the specific location of water sources and type of soils. Then, Subak farmers make together a deal in *Awig-awig* Subak about opening and closing the irrigation water supply that only applies to the WDUs (*tektek*). Therefor this routine is shown in Table 3.1 related to the social subsystem and the material subsystem, especially on the religious element. Also the system of one inlet and one outlet, as mentioned in Table 3.1 about the material subsystem on the natural relation element, enforced Subak irrigation schemes to apply the principle of justice in the management that has been adjusted to the natural environment and has proven to be an effective system. This effective system needs to be sustainable related to the agricultural productivity under limited water resources by updated operation and maintenance, based on a scenario analysis with five scenarios to improve agricultural production, that will be discussed in the next Chapters. However, the results can be accepted wisely by Subak farmers based on the element of *THK* on harmony among people.

Although the the Subak farmers are reluctant to change to new irrigation practices, such practices can in principle be useful for them as has been shown by Arsana (2012) and Sumiyati *et al.* (2013). The three scenario ideas by Lorenzen (2015) about: i) disintegration; ii) formalisation; iii) reinvention by contemplating the future of Balinese irrigation societies can inspire how the Subak communities, especially Subak farmers will adapt their practices in light of the future challenges.

At last, it is important that the Subak farmers believe and retain their indigenous knowledge in managing and maintaining the hierarchy of Subak temples as the right places to unite and persuade people in preserving the ecosystem.

4 Subak in the south of Bali: discharge analysis for a system approach to river basin development with Subak irrigation schemes as a culture heritage

4.1 Introduction

Depending on the location of the Subak irrigation scheme, an area of about 0.3 - 0.4 ha can generally be supplied by each water distribution unit. These units are supplied from the main/secondary/tertiary/quarternary canals within the system by using either continuous flow or an agreed scheduling of the water supply.

According to Geertz (1984), the details related to the technological elements of Subak irrigation schemes are very complex and have not been disclosed to researchers. In 1998 the Department of Public Works started to document the Subak irrigation schemes by addressing the traditional, technological and religious elements. This study was based on the understanding that the sustainability of indigenous paddy terraces depends on the availability of the discharge in the river at all stages in the river basin.

Therefore, one of the objectives of the present study was to analyze the available discharge from historical data on the weirs. This analysis can support the system of water supply to the paddy terraces in the Subak irrigation schemes in order to sustain agricultural productivity at all stages in the river basin.

4.2 Study of a river basin

The Yeh Ho River Basin (160 km^2) is located in the South of Bali. Yeh Ho River has a length of 45 km. There are 5,268 ha of irrigated fields along the river. Since the 1990s, the organization Subak Agung Yeh Ho is in charge of the management of this river basin.

Yeh Ho River is a perennial river. The basin characteristics of Yeh Ho River have an elongated shape with the main river on the right side (Figure 4.1). Yeh Ho River has three sections, which include upstream, midstream and downstream. The water diversion system in the upstream is called first time (*ngulu*), in the midstream second time (*maongin*) and in the downstream last time (*ngasep*).

Following source capturing of Gembrong Spring, the sequence of twelve diversion weirs is (Department of Public Works, 2004): Aya and Penebel (new weirs), Benana, Riang and Sigaran (old weirs), then Jegu and Caguh (new weirs) in the upstream, Meliling I in the midstream, and Gadungan, Sungsang I and Sungsang II in the downstream (Figure 4.1). In the period 2003 - 2006, Meliling II diversion weir was changed into Telaga Tunjung Dam in the midstream, built by the Department of Public Works. Due to the construction of the dam, the traditional way of water distribution has been influenced and therefore a new optimal distribution has to be determined. Because the topography of the river basin is relatively steep and the section of the river comparatively V-shaped, the storage volume is quite small compared to the storage height. The effective capacity of Telaga Tunjung Reservoir is 1 million m^3, and the height of the main dam is 33 m.

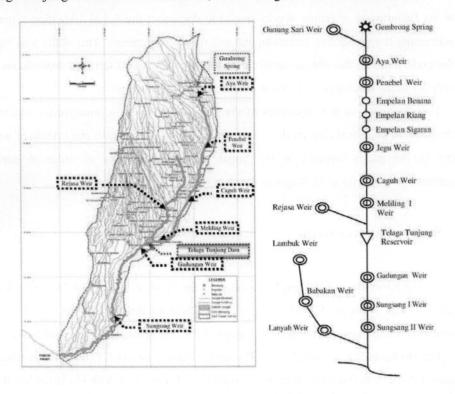

Figure 4.1. Yeh Ho River Basin and Subak irrigation schemes

4.3 Managed flow approach in Yeh Ho River Basin

In the Yeh Ho River the purpose of the weirs was and is to elevate the water level in order to enable gravity flow to the Subak irrigation schemes, and the purpose of the dam is elevating the water level and to store water as well. The present study can support the development of the arrangements of water supply to the paddy terraces of the Subak irrigation schemes in Yeh Ho River Basin in order to sustain agricultural productivity at the upstream, midstream and downstream level. Subak irrigation is especially taking place in Tabanan Regency. Due to the good soil fertility, Tabanan Regency successfully produced a paddy harvest at 22,455 ha of rice fields in 2010. The production was approximately 5 tons/ha of unhusked rice or 2.85 tons/ha of husked rice. Although the changes in land use from paddy fields to housing, farmlands and dry fields have increased with 0.3% over the last two years, the paddy production was relatively stable (Statistical Central Agency, 2010).

Based on the managed flow approach by Acreman (2010), the present study was conducted to determine the managed flow related to defined links between flow regime and function of the river, to define the managed flow options, and to assess impacts of the managed flow options. The results of the study may support the analysis of the river basin with the Subak irrigation schemes as specified in Figure 4.2.

4.4 Method and material

Gupta (2012) suggests that we need a paradigm shift towards *an information-based framework* for model identification, one that draws both on modern *system* and *information* theories, but also on the considerable conceptual (hydrological) knowledge that was historically developed and exploited before cheap computing made it so easy to analyse problems digitally.

Reservoir operation for water supply to Subak irrigation schemes in Yeh Ho River Basin

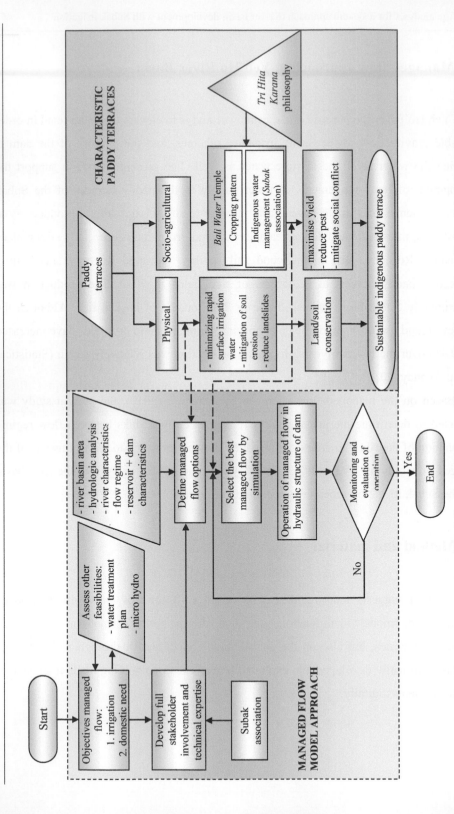

Figure 4.2. Development of managed flows model approach for sustainable water supplies to Subak irrigation schemes

Daily data inflows have been obtained from seven of the diversion weirs along the Yeh Ho River for which complete data of the period 2002 – 2010 are available. These data have been used for empirical flow frequency analysis. A popular method of studying the variability of streamflow is through flow duration curves that can be regarded as standard reporting output from hydrological data processing. The data were used for (DHV Consultants BV and Delft Hydraulics, 1999):

- evaluation of dependable flows in planning water resources engineering projects;
- evaluation of the characteristics of the hydropower potential of a river;
- assessment of the effects of river regulation and abstractions on river ecology;
- design of drainage systems;
- flood control studies;
- computation of the sediment load and dissolved solids load of a river;
- comparison of adjacent river basins.

The traditional way of water distribution of Subak irrigation takes advantage of mutual agreements on using water through the continuous flow system. Based on this system, the recoverable flow (*natak tiyis*) to the river also plays an important role as a source of water to the diversion weirs downstream. As a result, it is important to analyze the overflow weir data at the same time. The analysis has been done on a daily basis.

Meanwhile, in evaluating dependable flows, frequency analysis has been applied on the daily mean discharges. The analysis has been used in assisting the execution of the regulation and water distribution, which supplies the paddy terraces in the Subak irrigation schemes. Moreover, it has been used as inflow to the Telaga Tunjung Reservoir for further use downstream within the river basin.

By using the Weibull formula, the historic supply data of several diversion weirs were analyzed independently. This formula enables to provide a reasonably accurate forecast based on small samples, as well as a simple and useful plot, which is important to engineers and managers (Abernethy, 2002). In the Weibull formula the n values (number of years) are distributed uniformly between 0 and 100 percent probability, so there must be $n + 1$ intervals, $n - 1$ between the data points 1 and 2 at the ends (Chow *et al.*, 1988):

$$P(X \geq x_m) = \frac{m}{n+1} \qquad\qquad 4\text{-}1$$

where:

P   = probability plotting (%)

X   = random variable

x_m = probability distribution associated with the rank m

m   = ranking position

n   = number of years

The analysis procedure is as follows:

- the frequency or number of maximum to minimum occurrences *m* in *n* years of 365 daily discharge data has been selected. The daily inflow, diversion, and overflow data have been selected as well;

- the 80% of failure probability of the data has been analyzed to determine the minimum discharge. The percentage of probability of 50% can be considered as the mean discharge;

- the mean daily inflow, diversion and overflow have been plotted against the days;

- the water balance of each diversion weir has been represented by a graph.

4.5   Results and discussion

Before Telaga Tunjung Reservoir was built Subak Agung Yeh Ho managed 5,130 ha of paddy fields. Since the Telaga Tunjung Reservoir has been in operation the cropping patterns have changed as shown in Table 4.1. The cropping patterns in specific months have been determined and agreed by the Subak farmers. For many centuries, the cropping patterns as shown in Table 4.1 have been used and believed to be a fair system of water distribution. It is observed that the discharge in the river will remain the most important factor to sustain the cropping patterns. In addition to this the water balance of each weir (*empelan*) has been used to determine the availability of water in relation with supplying paddy terraces in the concerned Subak irrigation schemes.

Table 4.1. Cropping patterns in Subak Agung Yeh Ho before and after Telaga Tunjung Reservoir came into operation

Subak irrigation scheme	Functional paddy fields	Blocks			When to start land preparation
		Upstream (*Ngulu*)	Midstream (*Maongin*)	Downstream (*Ngasep*)	
	ha	ha	ha	ha	
Before					
1. Aya	644	644			Block I (*Ngulu*)
2. Penebel	731	731			Paddy I: Dec, Jan
3. Riang	25	25			Paddy II: July, Aug
4. Jegu	111	111			Block II (*Maongin*)
5. Caguh	1093		1093		Paddy I: Jan, Feb
6. Meliling	562		562		Paddy II: Aug, Sep
7. Sungsang	430			430	Block III (*Ngasep*)
8. Gadungan-Lambuk	1534		594	940	Paddy I: Feb, Mar Paddy II: Oct, Nov
Total field	5130	1511	2249	1370	
After					
1. Aya	644	644			Block I (*Ngulu*)
2. Penebel	731	731			Paddy I: Dec, Jan
3. Riang	25	25			Paddy II: July, Aug
4. Jegu	111	111			Block II (*Maongin*)
5. Caguh	1093		1093		Paddy I: Jan, Feb
6. Meliling-Timpag	142		142		Paddy II: Aug, Sep
7. Telaga Tunjung Reservoir					
• Meliling	420		420		Block III (*Ngasep*)
• Sungsang	430			430	Paddy I: Feb, Mar
• Gadungan	485		485		Paddy II: Oct, Nov
8. Lambuk	1187			1187	
Total field	5268	1511	2140	1617	

Source: Region River Office of Bali-Penida, 2006

The diversion graphs in the Figures 4.3a, 4.3b, 4.4a and 4.4b describe the probability of 80 and 50% of daily discharge patterns within a year for the two upstream weirs: Aya and Penebel. Similar trends can be noticed for the Q_{inflow} and the Q_{divert}. These trends occur during the dry season from June until October, when the wet season starts.

In the Aya Weir, the maximum diversions of 80 and 50% are respectively 0.63 and 0.71 m^3/s to supply the 644 ha of paddy fields. The minimum diversions are respectively 0.07 and 0.25 m^3/s. In the Penebel Weir, the diversions of 80 and 50% are respectively 0.85 and 0.87 m^3/s and the minimum divertions are respectively 0.18 and 0.36 m^3/s. These supply water to the 731 ha of paddy fields. The land preparation starts in December for Paddy I and in July for Paddy II. The fluctuating trends show that $Q_{overflow}$ of 80 and 50% occur from December until June. The peaks of Q_{inflow} of 80 and 50% in the Aya Weir are 1.12 and 1.47 m^3/s, while, the peaks of Q_{inflow} of 80 and 50% in the Penebel Weir are respectively 1.25 and 2.01 m^3/s.

In the midstream of Yeh Ho River there are the Caguh and Meliling weirs. In addition there is the Rejasa Weir in the tributary Yeh Mawa River as shown in Figure 4.1. The daily diversion graphs for the Caguh Weir and Meliling Weir at 80 and 50% probability are shown in the Figures 4.5a, 4.5b, 4.6a, and 4.6b. The months in which Q_{inflow} is fairly equal to Q_{divert} are from May, during the dry season, until November, when the wet season starts.

In the Caguh Weir, the diversions of 80 and 50% probability are 0.95 and 1.08 m^3/s. The minimum diversions are respectively 0.23 and 0.33 m^3/s. These provide water supply to the 1,093 ha of paddy fields. In the Meliling Weir, the diversions of 80 and 50% probability are 0.61 and 0.76 m^3/s, while the minimum diversions are 0.18 and 0.29 m^3/s. These can supply 562 ha of paddy fields. Land preparation starts in January for Paddy I and in August for Paddy II. The fluctuating trends show that $Q_{overflow}$ of 80 and 50% occur from December until May. In addition, the peaks of Q_{inflow} of 80 and 50% of Caguh Weir are 2.62 and 7.56 m^3/s, while the peaks of Q_{inflow} of 80 and 50% of Meliling Weir are respectively 2.23 and 4.31 m^3/s.

The diversions of the Rejasa Weir are shown in the Figures 4.7a and 4.7b. It has Q_{inflow} of respectively 2.28 m^3/s at 80% and 8.4 m^3/s at 50% probability. The minimum diversions at 80 and 50% probability are respectively 0.08 and 0.10 m^3/s.

Discharge analysis for a system approach to river basin development with Subak irrigation

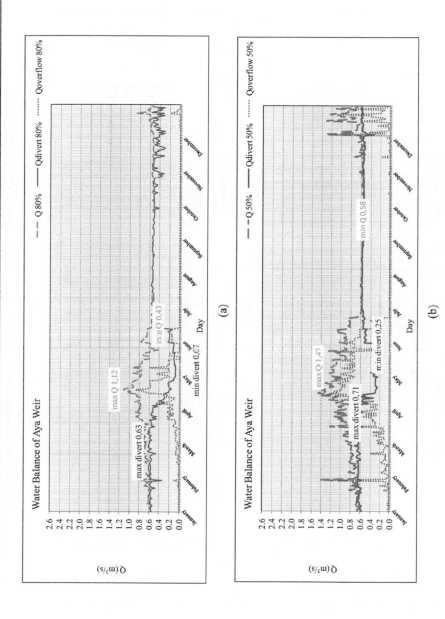

Figure 4.3. Daily flows over the Aya Weir at 80% (a) and 50% (b) probability

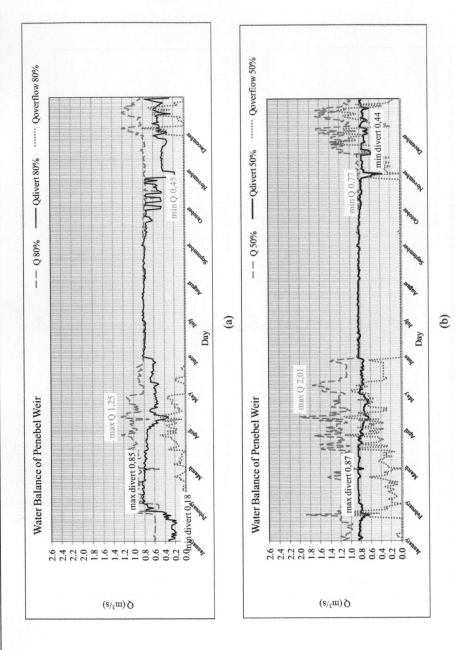

Figure 4.4. Daily flows over the Penebel Weir at 80% (a) and 50% (b) probability

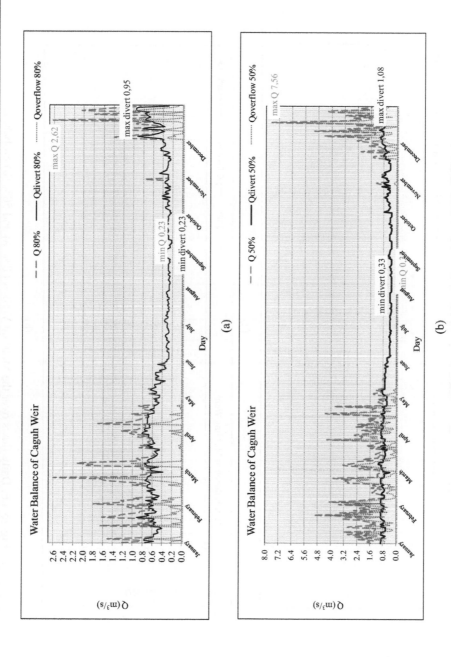

Figure 4.5. Daily flows over the Caguh Weir at 80% (a) and 50% (b) probability

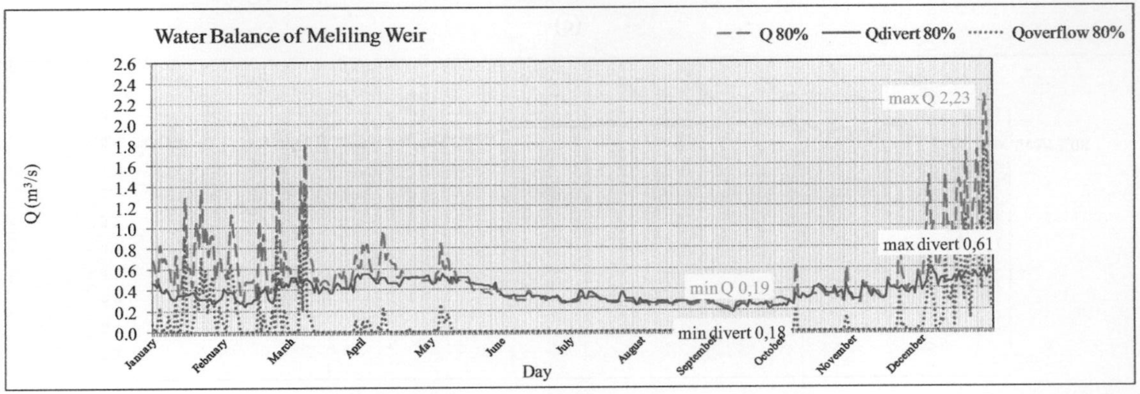

(a)

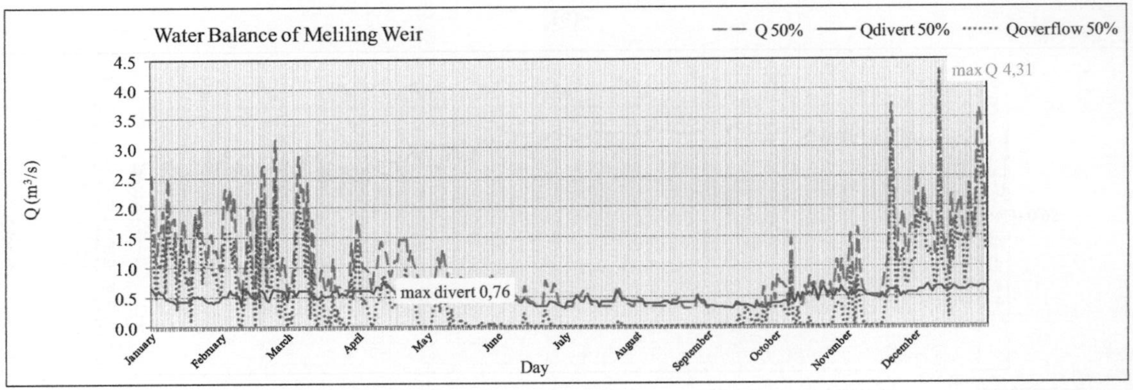

(b)

Figure 4.6. Daily flows over the Meliling Weir at 80% (a) and 50% (b) probability

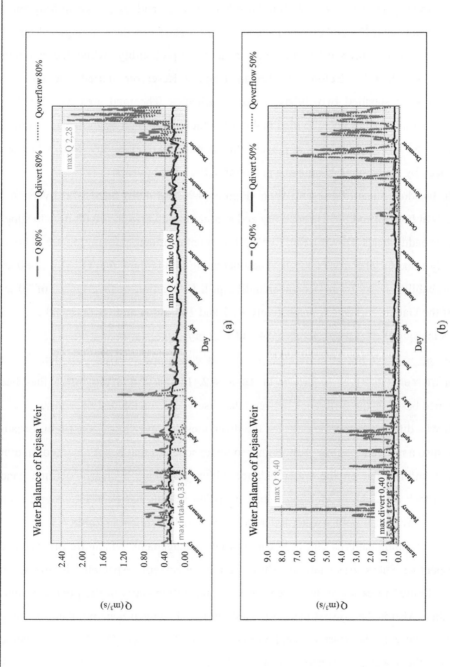

Figure 4.7. Daily flows over the Rejasa Weir at 80% (a) and 50% (b) probability

The diversions in the downstream Gadungan Weir at 80 and 50% probability are shown in the Figures 4.8a and 4.8b. Similar trends of Q_{inflow} and Q_{divert} start in May during the dry season until December when the wet season starts. In the Gadungan Weir, the diversion at 80% is 0.28 m^3/s, the same as at 50% probability, while the minimum diversion is 0.05 m^3/s. Before the Telaga Tunjung Reservoir started operation, the Gadungan Weir supplied two Subak irrigation schemes. The total area was 1,534 ha. Since 2006, this area has been divided in 485 ha, which is supplied by the Telaga Tunjung Reservoir and 1,187 ha, which is at present still supplied by the Lambuk Weir, which will be replaced by the Lambuk Dam and a reservoir in the coming years.

In the downstream Sungsang Weir (Figures 4.9a and 4.9b) there is a diversion of 0.61 m^3/s at both 80 and 50% probability and a minimum diversion of 0.06 m^3/s to supply 430 ha of paddy fields. Downstream, the starting months of land preparation are February for Paddy I and October for Paddy II. The fluctuating trends show that $Q_{overflow}$ of 80 and 50% probability occur from December until April. In addition to this, Q_{inflow} of 80 and 50% of the Gadungan Weir is respectively 4.31 and 16.0 m^3/s, and Q_{inflow} of 80 and 50% of the Sungsang Weir is respectively 5.84 and 17.9 m^3/s.

As an example the 50% monthly inflow, diversion and overflow for the Meliling section of Yeh Ho River is shown in Table 4.2. In this Table the inflow has been determined based on the difference between the observed diversion and overflow.

It was difficult to get the data for the diversions and overflows consistent, because there is also a recoverable flow from higher to lower schemes that is not recorded. In order to obtain information on the water balance within a scheme a clearly defined paddy terraces block of 4.7 ha in Meliling Subak irrigation scheme has been monitored. As an example the average monthly water balance for this block is shown in Table 4.3.

The cropping patterns and indigenous water management, which are organized by the respective Subak Associations, show in the midstream, especially in the Caguh Scheme, limited irrigation water when the land preparation starts in the period August - September. Also in the downstream, especially in the Gadungan Scheme, there is limited irrigation water at the start of land preparation in the period October - November. However, the Gadungan Scheme is supplied with sufficient water released from the Telaga Tunjung Reservoir.

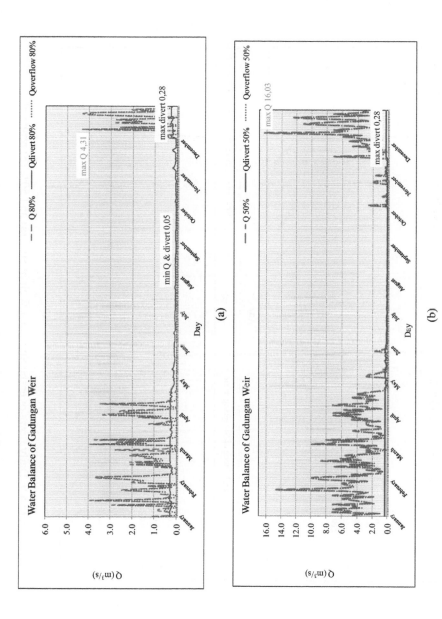

Figure 4.8. Daily flows over the Gadungan Weir at 80% (a) and 50% (b) probability

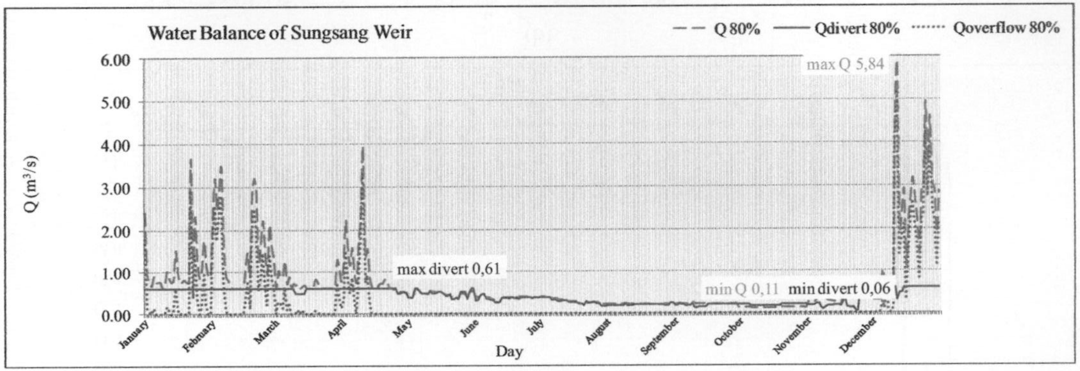

(a)

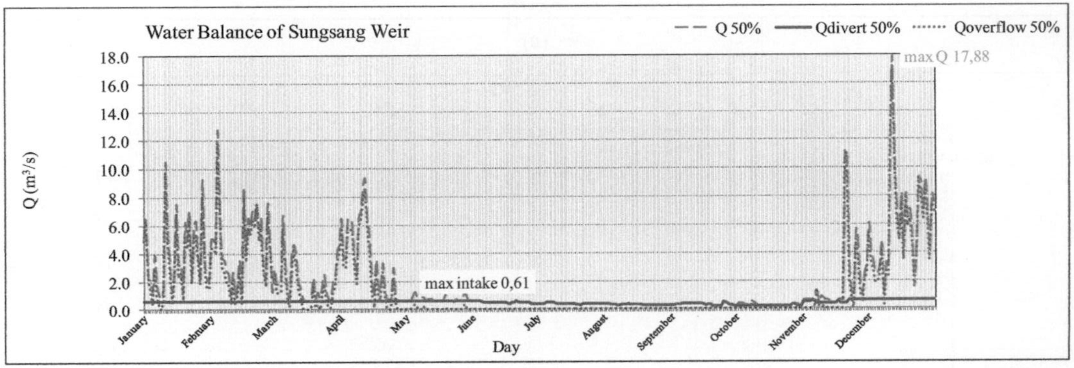

(b)

Figure 4.9. Daily flows over the Sungsang Weir at 80% (a) and 50% (b) probability

Table 4.2. Average monthly inflow, diversion and overflow of the weirs in the Meliling section of Yeh Ho River

	Diversion		Overflow		Inflow
	m^3/s	MCM	m^3/s	MCM	MCM
Jan	0.48	1.29	1.07	2.87	4.15
Feb	0.54	1.31	0.89	2.15	3.46
Mar	0.56	1.50	0.58	1.55	3.05
Apr	0.61	1.58	0.46	1.19	2.77
May	0.6	1.61	0.12	0.32	1.93
Jun	0.4	1.04	0.03	0.08	1.11
Jul	0.44	1.18	0.01	0.03	1.21
Aug	0.4	1.07	0.00	0.00	1.07
Sep	0.34	0.88	0.10	0.26	1.14
Oct	0.52	1.39	0.15	0.40	1.79
Nov	0.54	1.40	0.59	1.53	2.93
Dec	0.63	1.69	1.57	4.21	5.89
Annual		15.9		14.6	30.5

MCM = million cubic metres

4.6 Conclusion

In conclusion, the same trends of daily flows of all the weirs in the upstream, midstream and downstream show that the available water of Q_{inflow} can be diverted, because there is limited flow in the river during the period of June until October in the upstream, May until November in the midstream, and May to December in the downstream. Therefore, the concerned off-takes are kept open during those periods. This is indeed based on the agreement among the Subak farmers under the supervision of an irrigation observer who represents the Government, which proves the possibility of continuous flow throughout the water distribution unit (*tektek*) at the lowest level of the paddy terraces system.

Therefore, it should be noticed that the sustainable indigenous paddy terraces depend on the availability of the discharge in the main river that *needs to be sufficient for the diversions* from the upstream, midstream to downstream. This system would have to be ensured during the dry season, although in the wet season, the recoverable flow increases sharply to the downstream. The results of the study provide a perspective to *Subak* farmers on how to use the water more accountable.

As a result, the source of water in the upstream is extremely important to sustain the river system. This has become the main reason why 35% of the Gembrong Spring in the upstream is claimed by the Subak Agung Association. In contrast, most of the Gembrong Spring has been managed by the regency's water company to supply domestic needs. Then, it was a big challenge for the Subak Agung Association to supply sufficient irrigation water to the Subak irrigation schemes in Yeh Ho River Basin.

The hydrologic aspects of the dependable flows, while using trends in water balance discharge behind each weir in the river are very important to sustain the Subak irrigation schemes. The dependable flows of the upstream schemes, Aya and Penebel have to be considered (Figures 4.3 and 4.4), because of the sequence of irrigation water supplies based on upstream (*ngulu*), midstream (*maongin*), and downstream (*ngasep*).

Table 4.3. Average monthly water balances for a clearly defined paddy terraces block of 4.7 ha in Meliling Subak irrigation scheme

Average	WDUs/*Tektek*		Rainfall	Evapotranspiration	Recoverable flow		Additional supply
	m³/s	mm/day	mm/day	mm/day	m³/s	mm/day	mm/day
Jan	0.033	60.0	19	4.4	0.065	119	44.1
Feb	0.025	46.2	19	4.3	0.078	144	82.8
Mar	0.017	31.9	18	4.2	0.075	137	91.2
Apr	0.044	81.2	12	3.3	0.034	63	26.8 (storage)
May	0.032	59.5	13	3.2	0.069	127	58.4
Jun	0.036	66.7	12	2.7	0.116	213	137.9
Jul	0.024	44.8	15	2.7	0.088	163	105.5
Aug	0.022	39.9	2	3.5	0.075	139	100.1
Sep	0.028	51.2	5	4.0	0.030	56	3.3
Oct	0.027	48.8	4	4.3	0.026	48	0.7 (storage)
Nov	0.026	48.5	13	4.3	0.042	77	20.0
Dec	0.025	45.7	15	3.8	0.068	125	68.6
Annual in mm	18,700		4,400	1,340	42,300		20,500

5 Hydrology and hydraulic approaches: irrigation-drainage of Subak irrigation schemes, a farmer's perspective over a thousand years

5.1 Introduction

The irrigation and drainage arrangements for paddy terraces blocks have been applied by farmers using Subak's local wisdom for more than a thousand years. As long as they do not need the water through their fields, they seal the water distribution units using simple materials such as pieces of timber or clumps of soil. This system is called intermittent irrigation (*ngenyatin*). In addition, the drainage system has one or more drains, and or receiving water bodies (*pangkung*). The aim of intermittent irrigation and released drainage is to maintain an adequate amount of water in the paddy fields for weeding (*mejukut*) and fertilizing. In the intermittent irrigation period, the paddy roots can receive sunlight and conservation of water can directly take place.

Meanwhile, it is noteworthy that farming by consecutive irrigation is believed to have been applied during several centuries, as recorded in the inscription Sukawana (882) and inscription Bebetin (896) (Windia, 2013). Irrigation water is symbolically referred to as sanctified or 'holy water' termed *tirtha*. The notion of holy water as an ultimate means of purification and blessing is crucial for the understanding of the ritual control of the irrigation water and of the land, and the hierarchy of Subak temples that are the major institutions in charge of the ritual control of the flow of water. As one of the Subak temples (*Ulun Danu*) discussed in Chapter 3, the Batur Temple, as written on (only partly dated) palm leaf (*lontar*) manuscripts, was known collectively as the *Rajapurana* Batur. These manuscripts are kept in the Batur Temple and have been transcribed by Budiastra (1975, 1979). They outline the path of historical development of the institution (Hauser-Schäublin, 2011).

While in recent times Subak irrigation is facing water shortage and competition between different users the traditional way of irrigation and drainage arrangements has been influenced. However, in this situation *Subak Gede* along the river system still agreed to utilize the river water by scheduling based on the starting time of land preparation in the paddy cultivation seasons. Land preparation starts at the upstream blocks, followed by the midstream blocks and then by the downstream blocks. Regarding the cultivation schedule, the cropping pattern for each block is determined by mutual discussion.

In this study, a careful approach on the role of irrigation-drainage in Subak irrigation systems has been conducted by observing the water levels at the inlet and outlet of a paddy terraces block during two dry and two wet seasons. The combined result can support the efficiency of irrigation water supply to paddy terraces of Subak irrigation schemes, in order to sustain agricultural productivity.

5.2 Methodology

The focus of the study was on operation and maintenance of Subak irrigation schemes in conveyance of flow, especially in the role of irrigation-drainage based on the farmer's perspective. The simple constructions without sluice gates have enabled Subak farmers to: (i) regulate water based on a water allocation system agreed upon by all members; (ii) distribute water through the diversions to the paddy terraces blocks in varies ways, according to the availability of water simultaneously or rotationally; (iii) release excess water to drains. Therefore, members of a Subak Association have three activities as outlined in Chapter 3 (Yekti et al., 2013), of which the characteristics are:

- allocation and distribution of irrigation water to each member behind a water distribution unit (tektek/kecoran) with the following dimensions: 5 - 8 cm of width and 1 - 2 cm of height to a section of rice fields with an area of about 0.3 - 0.4 ha under a continuous flow system or an agreed scheduling system;

- the transfer mechanism among the members of the Subak Association behind a single weir within a Subak irrigation scheme, and between the Subak irrigation schemes. In the Subak irrigation schemes of Yeh Ho River the transfer of irrigation water from one scheme to another is regulated by mutual dealing among the water users and coordinated by the Subak leader (Pekaseh Subak Agung);

- the Subak Association has a mechanism for minor financing of physical improvement and maintenance at the lower level. The operation of the water distribution units is controlled by occasionally using intermittent irrigation. However, the water in the primary, secondary, tertiary and quarternary canals still flows to the lower schemes.

In the wet season when the water supply is sufficient, the intermittent irrigation is operated generally for 2 - 3 days three times in one cultivation season at the age of paddy of 10, 25 and 40 days. Thereafter there is no irrigation for 90 days up to harvest. At the same time the water is kept flowing continuously through the irrigation canals along the paddy terraces, together with the water release through the drains to the receiving water body (*pangkung*). The drainage water may become recoverable flow (*natak tiyis*) and provided as irrigation water to the downstream systems.

Subak farmers have the experience that continuous flow maintains the soil of the paddy bunds/levees at the required moisture content, thus cracking and sliding can be avoided. The paddy bunds/levees play various roles in the cultural landscapes, e.g.: water retention, footpath, source for hay to be fed to cattle, property boundary, space for landmark trees, drying space for harvesteed rice and space for dry foodcrops (Fukamachi *et al.*, 2005).

On the other hand, the water supply in dry conditions occurs during the long period of the dry season, as a result the water levels in the distribution boxes (*tembuku*) and in the water distribution units (WDU) may become low and insufficient to supply water. Due to this intermittent irrigation hardly can be applied and there is limited, or no release of drainage water from the paddy terraces. Furthermore, Subak farmers can flush irrigation water from the Subak irrigation systems in the upstream, in order to meet the required water levels using the transfer mechanism, which was explained in the previous chapter.

A scientific perspective on farmer's practices needs to be based on observations in paddy terraces blocks, in the distribution boxes and in the water distribution units, as well as in the drains from the paddy terraces. Therefore, quantitative data have been collected on the role of irrigation-drainage within a paddy terraces block of 4.7 ha in the Meliling Subak irrigation scheme.

5.2.1   Observation of the water balance in a paddy terraces block

In order to analyze the crop water requirement in the paddy terraces block the discharge from diversion canals (quarternary box) as beneficial consumption (Perry, 2007) and recoverable flows (drain/*pengutangan*) have been observed within the sample paddy terraces block as shown in Figure 5.1. The collected data have been analyzed to provide the trends of water requirement during a cultivation season.

Inlet by quarternary box

Outlet from paddy terraces block

Figure 5.1. The observed paddy terraces block in Meliling Subak irrigation scheme

5.3 Results and discussion

The application of continuous flow in the Subak irrigation schemes has been investigated in detail based on observations and analyzing the discharge data of several weirs (*empelan*) to recognize trends in the river system (Yekti *et al.*, 2012). Subsequently during this study daily flows have been observed and analysed based on the farmer's practise with the water distribution unit. The observations of height, width and velocity of water at the inlet and outlet of the paddy terraces block (Figure 5.2) have resulted in the trends of daily flows in the distribution boxes, water distribution unit and drain (irrigation-drainage).

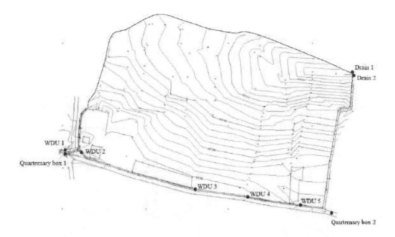

Figure 5.2. Lay out and observation points of the observed paddy terraces block in
Meliling Subak irrigation scheme

Two components of the *THK* philosophy, harmony of people and nature and harmony among people, have been implemented by the Subak farmer's during the period of observation from April 2013 until April 2015. In 2013 during the early phase of the dry season from March until April it was still raining. Although in the observed scheme and in other midstream schemes, the Subak farmers would have to follow the midstream block cultivation season, which starts in October - November 2013, they started to plant the paddy crop based on the natural conditions by mutual decision in May - June 2013 (5 months earlier). This was followed by a fallow period from September until mid December 2013. Thereafter they started to plant the paddy crop by the end of December. Harvesting took place in April 2014, within the wet season. After harvesting, the fallow season from April 2014 until the middle of November 2014 was used for improvement of the primary and secondary canals, while the reservoir gate for irrigation was closed. Thereafter the reservoir gate for Meliling Subak irrigation scheme was opened again.

The observed irrigation-drainage flows during the two years showed that the Subak farmers use the irrigation water and drain the excess water in a natural way during the cultivation seasons, as shown in Figure 5.3a (period April 2013 - April 2014) and Figure 5.3b (period April 2014 - April 2015). The irrigation water was used quite effective with supplies between 0.02 m^3/s and 0.04 m^3/s.

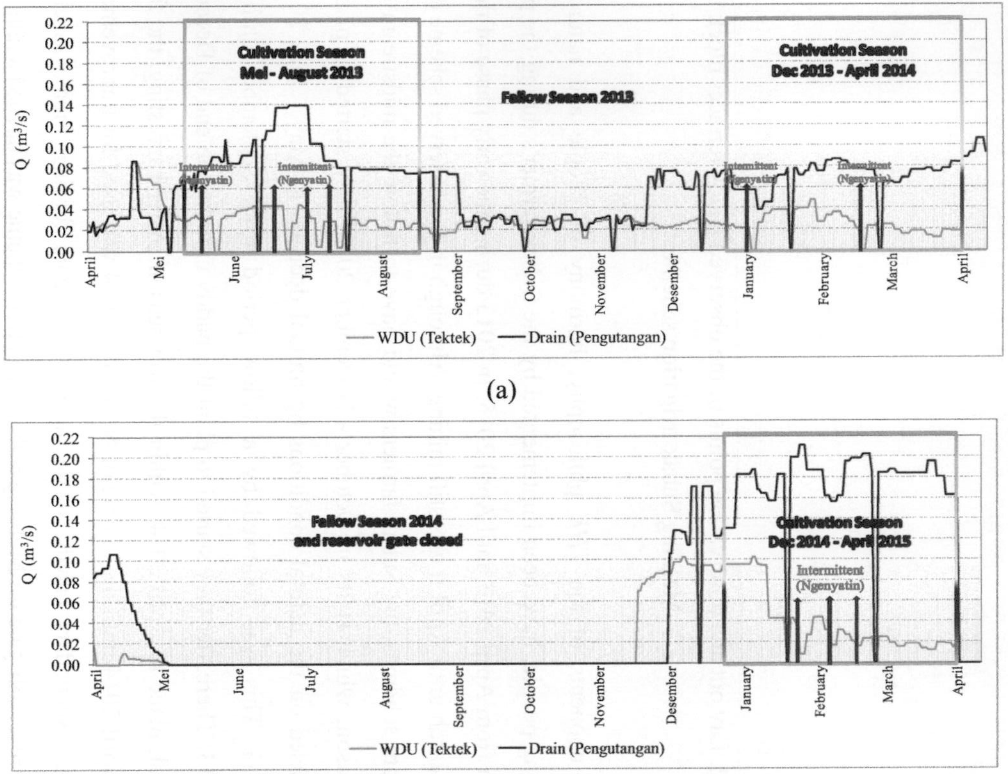

(a)

(b)

Figure 5.3. The irrigation-drainage flow observations in the WDU and drain of the observed paddy terraces block in Meliling Subak irrigation scheme

The drainage water increased sharply with values of 0.075 - 0.14 m^3/s in the first season and 0.04 - 0.08 m^3/s in the second season.

In the second year, the irrigation water supplies were for the same paddy terraces block between 0.02 and 0.09 m^3/s higher than in the first year. The drainage water was quite high during one cultivation period, between 0.14 and 0.21 m^3/s.

The water level at the five water distribution units with varying heights and widths resulted in the same trends (Figure 5.4). The intermittent irrigation was conducted four times in the first cultivation season, two times in the second and three times in the last (Figure 5.3), while the land of the paddy terraces block had received enough rainwater during the first cultivation season. Also during the last cultivation season the rainfall was quite high. The water level in the drains increased sharply during the first cultivation season and in the period November 2013 - January 2014 when in the second cultivation season the land preparation was done (Figure 5.5). The same trend was observed in the last cultivation period from December 2014 - April 2015.

In spite of the continuous flow quarternary box 1 had enough water with discharges of 0.02 - 0.06 m^3/s and 0.02 - 0.10 m^3/s to irrigate the paddy terraces block as shown in the Figures 5.6a and 5.6b. From 24 to 29 January 2015 the canals were closed for maintenance. However, the Subak farmers did not plant the paddy crop and preferred to conserve their paddy fields during the fallow period. In addition, there is conveyance flow from one block to a downstream block, which is shown in Figure 5.6. The continuous flow of quarternary box 2 has varying discharges to supply irrigation water.

Recently a research was done by Arsana (2012) on the water requirement for paddy cultivation in relation to the season, location, irrigation and varieties of paddy in the upstream, midstream and downstream sections of Yeh Ho River Basin. This study explained that irrigation during 8 days gave a yield of paddy of 1.24 kg/m^3 while continued irrigation gave 1.14 kg/m^3. Moreover the application of the System of Rice Intensification (SRI) combined with intermittent irrigation was investigated by Sumiyati et al. (2013). This study was done to determine the effect intermittent irrigation (ngenyatin) and the SRI method on rice productivity in the paddy field of Subak Sigaran, Tabanan Regency. The results show an increase in productivity of 18.4%, as well as increase of tillers, of the length of panicles and of total grains per panicle.

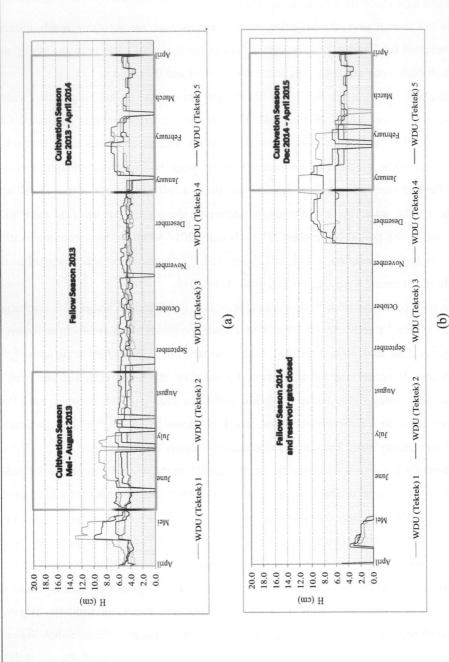

Figure 5.4. Profile of the water levels at five WDUs in the Meliling Subak irrigation scheme by daily observations

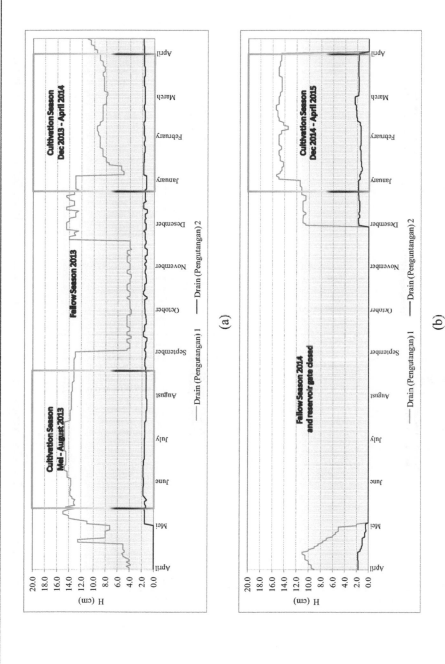

(a)

(b)

Figure 5.5. Profile of the water levels in two drains in Meliling Subak irrigation scheme by daily observations

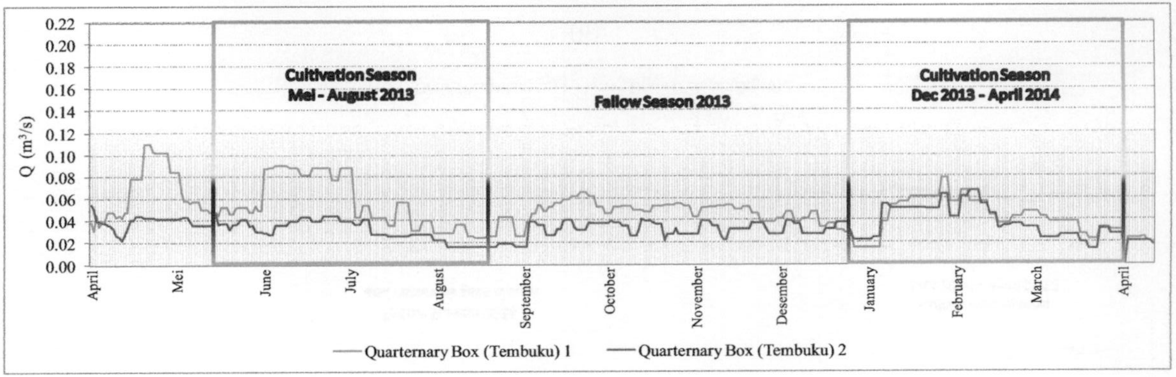

(a)

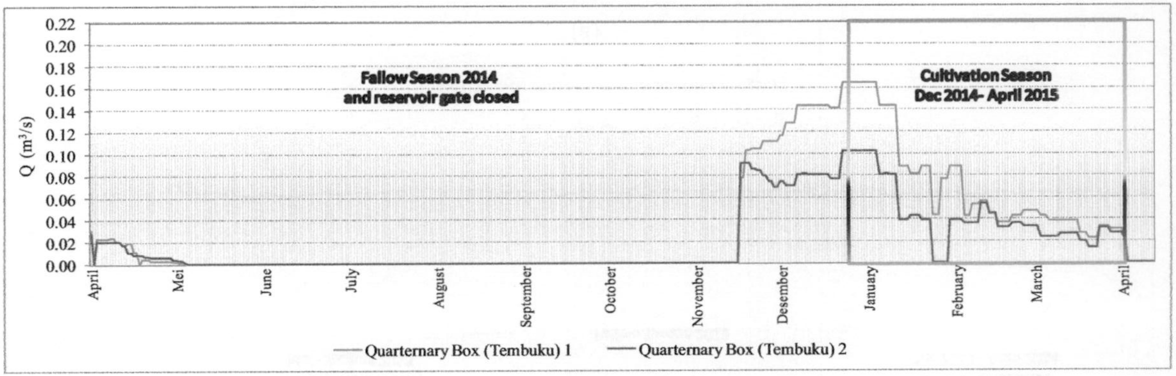

(b)

Figure 5.6. Daily flow observations in two quarternary boxes in Meliling Subak irrigation scheme

However, there is a problem to transfer these worthy results to the Subak farmers in order to encourage them to operate and maintain the paddy terraces fields in the new way, while in the farmer's perspective the SRI method is not easy applicable in practice for reasons such as:

- the seeds should be developed to sprouts, thus in the SRI method the 7 - 12 days developed seeds need to be planted more carefully than with the traditional method, which has 15 - 20 days developed seeds. Also with the SRI method the seeds need to be planted by using a high precision technique (the length and the total seeds of a panicle), which needs 1 - 2 days more than the traditional method by employing five persons/hectare. In reality now, the number of people who work in the agricultural sector decreased and there is limited experience with the SRI method, sometimes the Subak farmers have to pay a person to help them, or they employ their family and neighbours, who have limited knowledge about the SRI method;

- for strengthening of the grains the SRI method needs a dry paddy field during 7 days after seeding. On the other hand, the traditional method does not need a dry paddy field. In addition, there is the farmer's reason that the continuous flow through the paddy fields can avoid the growth of weeds and seeds of weed. This can result in diseases in future. In this period, the Subak farmers will implement the intermittent irrigation for clearing of weeds during 2 - 3 days at the age of paddy of 10 days;

- the Subak farmers had experience with the SRI method in the paddy terraces fields during the cultivation season. As a result of this practise cracking of the paddy bunds/levees occurred with development of cracks in the paddy fields. Due to this the next land preparation was difficult;

- nevertheless, the Subak farmers agreed that 15 days before harvesting they will dry the paddy terraces field.

5.4 Conclusion

Initially the *THK* philosophy has been implemented by the Subak farmers in the investigated paddy terraces block in Meliling Subak irrigation scheme, who made the right decision to start land preparation based on their indigenous knowledge about the signs of the start of the wet season and their mutual discussion. The results of the study show a conformance of justice among nature, human and technology that has been applied by using an orderly system for adjusting the topography, soil and water sources, mainly rain, which became the important values to operate and maintain the paddy terraces block, as well as to manage the river basin. Because of this continuous flows in quarternary canals were used in an effective way and the excess water was released and where applicable directly used as recoverable flow (*natak tiyis*) for downstream schemes. The irrigation-drainage process has shown that conservation of water has been implemented during the concerned periods by considering the condition of the hydro-climate in the river basin. This concerns proper water distribution and allocation management during the dry season as well as efficient drainage management during the wet season, to meet in an optimum way the soil moisture demands of the crops in the root zone during both seasons. It is also expected to provide a quantitative description on how to manage water in Yeh Ho River Basin.

Finally, an important perspective would have to be considered by the Subak farmers, who are competing with other users, to be aware that irrigation water can be more valuable. Thus, it will be interesting to elaborate on this study towards water pricing in Subak irrigation schemes, even though the two preceding worthy Subak studies related to the growth of rice productivity and water requirement were conducted in recent times. Therefore, the variety of sciences of Subak studies will become a valuable source for recommendations to the Government. In the next Chapters, based on river basin simulation the quantity of water, productivity and pricing in the Subak irrigation schemes will be discussed.

6 Model simulations and optimisation technique

6.1 Model categorization

'As modellers or researchers, we must discipline ourselves to work more closely with our clients - the planners, managers and other specialists who are responsible for the development and operation of our water resources systems. We must study their systems and their problems, and we must identify their information needs' (Loucks, 1992; Hejazi and Cai, 2011).

Modelling and analysis methods for evaluating the water supply capabilities of reservoir/river systems are fundamental to the effective management of the highly variable water resources of a river basin (Wurbs, 2005). The modelling would have to be supported by scenarios that are self-consistent story lines of how a future system might evolve over time in a particular socio-economic setting, and under a particular set of policy and technology conditions. The scenarios are formulated and then compared to assess their water requirements, costs and environmental impacts (Mugatsia, 2010).

In order to achieve the best decision a system analysis can be effectively applied. System analysis is an efficient and systematic rational approach used to gain the best decision of a system in which its approach is based on the available information and its limitations (Sudjarwadi, 1992). There are two common techniques used in the analysis namely simulation and optimisation. A simulation technique is a quantitative method that describes the behaviour of a system, which is used to estimate system outputs. Different simulation and optimisation techniques have been developed and implemented to incorporate the uncertainties, such as climate variations and variations in market prices, which have caused uncertainty on agricultural water demands in water management policies (Zahraie and Hosseini, 2009).

Several models have been developed for simulation and optimisation in irrigation water resources, such as an integrated optimisation method using Stochastic Programming (SP) that was developed for supporting agriculture water management and planning in Tarim River Basin, Northwest China (Huang *et al.*, 2012). The hydrological model is aplied for forecasting the available irrigation water and the simulation system is then embedded into an optimisation framework, where the objective is to maximize the system benefit for water resources management.

Linear programming has been used to optimise reservoir operation since many years. Rani and Moreira (2010) and Alemu *et al.* (2011) highlighted some of the advantages of using linear programming. These include flexibility for application to large-scale problems, convergence to global optimal solutions, and the wide variety of commercial and open-source software packages available. In addition, the mathematical models behind linear programming and their objective functions can be relatively straightforward explained to stakeholders.

Recently mathematic programming based on genetic algorithms has been developed. Genetic algorithms are optimising algorithms, inspired by natural evolution. They produce a complete population of answering points. Each point is tested independently to establish new populations, including modified points (Dehini *et al.*, 2012; Kiyoumarsi, 2015). Genetic algorithms consider many points simultaneously and these characteristics conform them with more parallel processors, since considering that each point needs some calculations, including target functions differentiation, etc. (Sivanandam *et al.*, 2007; Kiyoumarsi, 2015). In these algorithms, different operands and mechanisms are implemented, that are described here to analyse the applicability of genetic algorithms as optimising algorithms (Asfaw *et al.*, 2011; Kiyoumarsi, 2015) mainly for natural evolution simulation and only to a certain extent based on strong mathematical theories. This technique uses natural genetics, statistical methods (Kiyoumarsi, 2015) and a type of creativity, and tries to reach an optimised conclusion (Kiyoumarsi, 2015).

The genetic algorithms model gives better results when compared to the linear programming model on reservoir operation for known total irrigation demand and on the optimal allocation of water available to crops at the farm level (Md. Azamathulla *et al.*, 2008). Moreover, genetic algorithms are well suited to solve irrigation scheduling problems. They are robust and very efficient and can easily be run with a range of objective functions. The developed genetic algorithms could with little difficulty be applied to problems that are more complex.

In this study, the simulations have been conducted based on the diversion by the weirs in the Yeh Ho River and on the storage in Telaga Tunjung Reservoir to supply the

Subak irrigation schemes with the required amounts of irrigation water in a timely and efficient manner, based on the cropping patterns. The physical characteristics of the Subak irrigation schemes and their operation rules can be modified to adapt to the changing circumstances. For example, sustainability has been incorporated in irrigation water management, which can be defined as managing irrigation water in an economically efficient and socially equitable manner, while considering conservation of the environment (Kang and Park, 2014).

6.2 Modelling of Subak schemes related to *Tri Hita Karana* philosophy

The modelling has been based on the technological and socio-agricultural approaches. The boundary conditions were based on the local conditions formed by religious elements and the Balinese water temple. The religious elements of *THK* have inspired the application of the local cropping patterns and indigenous water management. The possible effect of the Telaga Tunjung Reservoir on the water supply to the Subak irrigation schemes in Yeh Ho River Basin has been analysed. The irrigation plays an important role in agricultural livelihood strategies that support and strengthen the cultural values of the indigenous people (Groenfeldt, 2005), as well as to further develop their agricultural products.

Before describing the modelling and scenario analysis of the reservoir operation in connection with the water supply to the Subak irrigation schemes, the conceptual approach of the research has been defined based on the literature review and the description of the study location as shown in the Figures 4.1 and 4.2. The *THK* as principal value is underlying the reservoir operation and the supply to the Subak irrigation schemes. This is why the *THK* triangle acts as a boundary condition (Figure 3.2). However, the *THK* philosophy is not an accountable value that can be put in the reservoir operation analysis, but it can give the belief to act based on the paradigm how to analyze, simulate and evaluate the result.

The role of reservoir operation based on the scenario analysis and simulation process of the Subak irrigation schemes in Yeh Ho River Basin is included in the content material subsystem of point 3 of the *THK* philosophy as shown in Table 3.1.

6.3 Multiple purpose reservoir operation

The first purpose of Telaga Tunjung Reservoir is to serve the Subak irrigation schemes with water up to 1.87 m³/s. It concerns 420 ha Subak Meliling, 485 ha Subak Gadungan and 430 ha Subak Sungsang. The second purpose is to deliver 120 l/s raw water, which is divided into 24 l/s for industrial centres and domestic needs in the Village of Berembeng, 50 l/s for the development of Soka tourism area, and 46 l/s for domestic water in the Selemadeg and Kerambitan districts. If these purposes have been fulfilled, other purposes can be developed, such as domestic needs, micro hydropower (micro-hydro) and fishing that approximately uses 10% of the reservoir surface area. Starting in 2013, the purposes for domestic needs are in operation, but the micro-hydro is not yet in operation.

Since the impounding process of the reservoir began in December 2006 the operation was observed from January 2007 to December 2012. The operation is based on the existing cropping pattern, but the operation rule, which can provide optimal agriculture production has not yet been determined.

6.4 Scenario analysis in Subak schemes

As shown in Figure 6.1, the technological and socio-agricultural elements, the application of local cropping patterns and indigenous water management practices are regarded as the boundary conditions, which will be translated into scenario's of simulation at river basin level. In the Subak schemes, the cropping patterns, which fall in specific months, have been determined and agreed upon by Subak Agung Yeh Ho, as shown before and after the initiation of the Telaga Tunjung Reservoir in Table 4.1.

While the period of land preparation is critical with respect to the water needs the scenario analysis and optimisation have been based on shifting of the starting time of land preparation (*nyorog*). The objective function of the optimisation is to obtain an optimal agriculture production for all schemes in Yeh Ho River Basin. The simulation and optimisation have been analysed according to Figure 6.1. Details of the scenario analysis and the results will be explained in section 7.4.

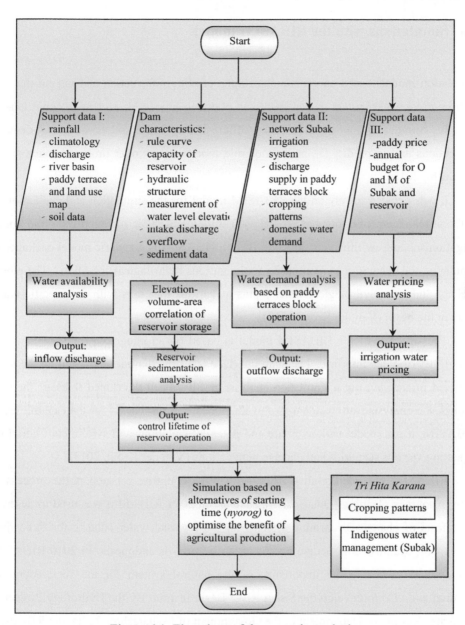

Figure 6.1. Flowchart of the scenario analysis

6.5   Simulations with the RIBASIM model

Simulation models based on genetic algorithms provide information and insight that can help improve water system management and planning processes, such as in conditions of drought. Simulation models provide an efficient way to reproduce source - demand interactions and to predict the impacts of rule modifications over time and space (Sulis and Sechi, 2013).

It was decided to rely on the RIBASIM model for the simulation and evaluation of different management options of the Telaga Tunjung Reservoir and supplies to the Subak irrigation schemes within Yeh Ho River Basin. RIBASIM is a generic model package for simulating the behaviour of river basins under various hydrological conditions. The model package links the hydrological water inputs at various locations with the specific water users in the basin (Van der Krogt, 2013).

The structure of the RIBASIM model is based on an integrated framework with a graphically oriented interface. The main interface is a flow diagram representing the tasks involved in carrying out a simulation analysis and has been developed to assist the user through the analysis from data entry to evaluation of the results. As for the capabilities of RIBASIM, it can model various future and potential situations and system configurations by setting various scenarios and management actions (Van der Krogt, 2013).

RIBASIM has been applied in more than 20 countries to support the process of water resources planning (Omar, 2014), such as: in 2001 RIBASIM was used to describe the effects of changes in farming systems on the regional water balance for three river basins - the Jratunseluna, Serayu and Cidurian basins - in Indonesia. In 2010 RIBASIM was applied to clarify the importance of the natural system for the socio-economic situation and also to develop the Sistan River Basin in Iran. By the Hydrology Project of the Water Resources Department of the Government of Maharashtra, India, the RIBASIM model was applied to predict the water shortages in Godavari River Basin, India for 2015 and 2020, and to develop options for minimizing water deficit.

6.6 Application aspects of the RIBASIM model

The simulation with the RIBASIM model proceeds in time steps (Figure 6.2). For each time step the water balance is computed based on the supply of water at the boundaries of the system, the demand of water by the various users, the operation rules for the various structures, such as surface water and groundwater reservoirs, weirs and the water management policies at the basin level. The water diversions are based on the operation rules of the hydraulic structures, as well as on the cropping pattern, based on the month when land preparation starts. This created the scenarios for the simulation process.

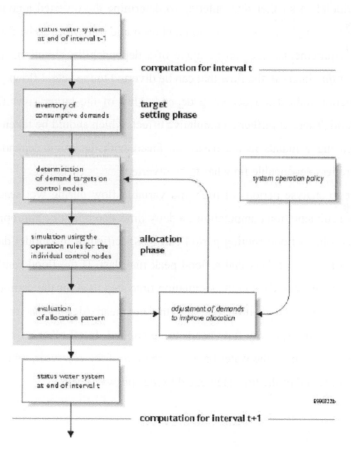

Figure 6.2. Phases in a time step in the RIBASIM model (Van der Krogt, 2013)

The time step to represent a water balance needs to accommodate the variations in the sources of supply and demand. Variations from time to time in demands can usually be represented adequately by a half-monthly or monthly interval. However, the hydrologic variations in rainfall or flow can vary widely.

Quite frequently different interlinked supply - allocation systems in a river basin require a different time interval and a compromise solution is required. Typical for this case is the supply of a particular demand from the reservoir and a regulated diversion.

For the storage system, a half monthly or monthly time interval will be appropriate to describe variations and allocation decisions. The diversion from an unregulated river with a relatively fast varying flow (e.g. strong daily variations in response to rainfall) requires in principle a smaller time interval to determine the diverted amounts of water. For comparison, a single interval needs to be chosen and a way has to be found to match the different requirements. Use can be made of a dependability function, which relates, over a half monthly interval, the flow that can be diverted to the actual flow.

The function indicates a declining dependability of the larger river flow above a certain threshold. There is further a cumulative effect, which should be taken into account because increasing demands in the basin are faced with declining dependable flow as more and more less dependable flow has to be diverted.

In order to analyse periods of high and variable flow during wet seasons in more detail, e.g. for sedimentation computations, a daily time step might be appropriate.

For reservoirs, a flood routing period can be specified in the model data reflecting the duration in number of days that a flood peak may last. If the flood routing period is bigger than the number of days in the simulation time step then for the computation of the spillway release flow the relation between the net head (defined as the head above spillway level) and the spillway flow needs to be applied. In all other cases, the spillway flow is computed based on the water balance equation of the reservoir, which results in a storage level at the end of the time step equal to the spillway level.

6.7   Yeh Ho River system as input in the RIBASIM model

According to Ali and Shui (2001), efficient utilization of water resources needs information data, such as, annual effective rainfall, runoff, consumptive use and reservoir release. Effective processing of these data is possible through a state-of-the-art computer modelling system. Thus, a reservoir operation rule is to be developed to meet shortage of water to some extent. The schematisation for Yeh Ho River Basin is shown in Figure 6.3.

As the first step, adequate and reliable data have been collected. Highly reliable data are important since they contribute to any settlement of the research efforts. The required data that were collected consist of primary and secondary data.

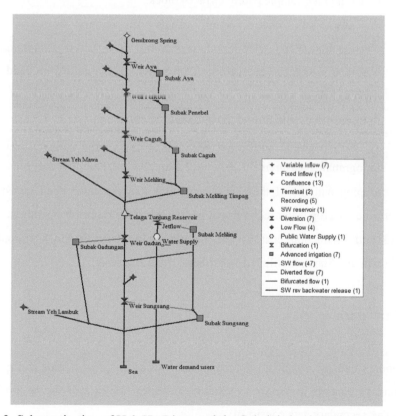

Figure 6.3. Schematisation of Yeh Ho River and the Subak irrigation schemes in the river basin in the RIBASIM model

Primary data

Analytical modelling tools constructed and evaluated based on primary data can help assess effects of alternative operational scenarios related to reservoir and Subak irrigation scheme operation, water right transfers and changes in irrigation practises (Berris et al., 1998)

Primary data have been collected from surveys, which give information about the existing Subak irrigation schemes and the existing reservoir operation. Paddy terraces have physical and socio-agricultural systems, which have been followed in this research. Furthermore, collected primary data have been entered into the RIBASIM model:

- these concern in the sample paddy terraces block:
  * the discharge from diversion canals (quarternary box) as beneficial consumption (Perry, 2007) and recoverable flows (drain/pengutangan) were observed in order to analyze the crop water requirment of the paddy terraces blocks;
  * the geologic profile to find the soil parameters;
  * by in situ tests infiltration rates have been determined;
  * topographic measurements have been done to observe the elevation.
- the cropping patterns of the Subak irrigation schemes have been investigated to determine the crop water requirement in relation to the reservoir operation;
- yield and production cost of the sample paddy terraces block (Subak Caguh upstream and Subak Meliling downstream of the reservoir).

Secondary data

The secondary data are needed to support the analysis of design standards and approaches. The secondary data are important to generate the operation rule of the reservoir and its lifetime. The secondary data consist of:

- hydrologic and climatologic data such as annual effective rainfall, evapotranspiration, humidity and sun intensity;
- map of the river basin;
- discharge records of the weirs in the main river;
- physical data and design of the reservoir;
- land use in the river basin;
- soil data;
- records of water levels, seepage discharge and relief wells, intake, outflow and overflow;
- domestic needs;
- reservoir sedimentation;
- harvested paddy prices.

7 Scenario analysis

7.1 Hydrologic and hydraulic analysis

River system modelling and analysis related to reservoir operation need hydrologic and related data. Long records representative for historical hydrology are used in planning studies. Existing observations are required for real-time operations. Moreover, a selection of computational tasks is involved in converting records of field measurements on hydrologic processes to the input data required for river basin development and reservoir system analysis models. In light of this, historical records of gauged streamflows need to be adjusted to represent flows at relevant locations for a specified condition of river basin development. For instance the analysis of weir streamflow data in Subak irrigation schemes within Yeh Ho River Basin. The results can give an idea how the water management in this river basin can be developed.

River and reservoir system analysis models are built around water accounting and routing algorithms. Water accounts are balanced as water is routed through the system. In a reservoir operation model, the streamflow provides the inflow to the system. Reservoirs regulate the streamflow through storage and releases. Water flows through river reaches, can be diverted for beneficial use, which can result in evapotranspiration, and there can be other gains and losses.

A river is a natural condition of open channel hydraulics. Open channel hydraulic models can simulate flow conditions in natural and improved streams, associated floodplains and constructed canals. Water surface profiles are needed for various reservoir operation rules to release water for related water management applications. Intake structures may become inoperative if river stages drop below certain levels. Evaluation of flood control operations and floodplain management programs, design of levees and channel improvements based on water surface profile models, erosion and sedimentation may be of significant importance in design and operation of river control structures. Flow rates and velocities computed with hydraulic models can provide basic input required by water quality models.

7.1.1   Analysis of rainfall data: dependable rainfall and effective rainfall

Rainfall data concern one of the parameters in a hydrologic analysis for reservoir operation. The dependable and effective rainfall are used for reservoir simulation and consumptive use of irrigated crops. The dependable rainfall is the amount of rainfall that can reliably be expected during a certain period in an area where the risk of yield reduction, or even failure plays a role. In Indonesia, the standard for dependable rainfall for irrigation is usually set at 80%. Dependable rainfall data can be analyzed with the Weibull formula as shown in equation 4-1. In Yeh Ho River Basin, dependable rainfall in the middle range (101 – 300 mm/month) occurs from November until February. Thereafter heavy rain, occasionally upto 300 mm/month, can occur during March and April (Annex C). The Subak farmers take the decision to start land preparation (Paddy I) based on their indigenous knowledge about the signs of the start of the rainy, or wet season. They use this capability based on the months of the wet season. In the Subak irrigation schemes, they have three parts in sequence (Table 4.1): Block I (the first/*Ngulu*) Paddy I: December, January; Block II (the second/*Maongin*) Paddy I: January, February; Block III (the last/*Ngasep*) Paddy I: February, March. The cropping patterns illustrate the equity of the schemes within the river basin.

Effective rainfall has various definitions from the point of view of crops, soil moisture, management of irrigation systems, and economics (Chen *et al.*, 2014). Hayes and Buell (1955), Li (1956), Chin (1965), and Dastane (1974) stated that effective rainfall is the quantity of rainfall available for plant growth. Those who consider just the water available for plants, Tsao (1966) and Chen (1979), indicated that the conjunctive use of effective rainfall can reduce the amount of irrigation water from rivers and reservoirs. Ogrosky and Mockus (1964) defined effective rainfall by considering the water that enters into the soil. Hershfield (1964), Lo (1962), United States Department of Agriculture (USDA) (1967), Jackson (1992), Obreza and Pitts (2002), Liu *et al.* (2007), Rahman *et al.* (2008), and Adnan and Khan (2009) defined effective rainfall by considering the amount that just meets the consumptive water requirement.

The effective rainfall is used for efficient consumption of water in irrigated agriculture, such as for design and operation of irrigation and/or drainage systems, for rainfed agriculture, or for leaching of salts. Other aspects that have an influence are: temperature, solar radiation, humidity, wind movement, length of the growing season, latitude and sunlight.

In this research, the rainfall data were taken from Kerambitan Station. The effective rainfall has been determined by using the empirical equation for irrigation of rice of the Irrigation Planning Standards (*Standard Perencanaan Irigasi*) and the Planning Criteria (KP) 01 by the Department of Public Works. The effective rainfall is taken by 70% of the partly-month (15 days) rainfall with 80% probability (Figure 7.1; Annex D):

$$R_e = 0.7 \times R\,80_{(partly-month)} \qquad\qquad 7\text{-}1$$

where:

R_e = effective rainfall (mm/day)

R80 = 80% probability of dependable rainfall (mm/day)

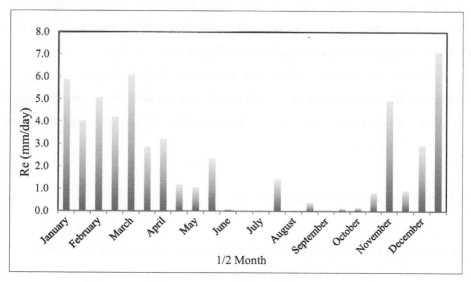

Figure 7.1. Effective rainfall for paddy terrace schemes

7.1.2   Streamflow analysis

Since the future is unknown, the historical streamflow needs to be used as a representation of the hydrologic characteristics of a river basin to be expected during the future planning scope. A planning study may involve analyses of the operation of existing reservoirs and/or evaluation of proposed new projects. Evaluation of proposed long-term improvements in operation policies may be of concern, or the objective may be to formulate operation strategies for the next season or year.

Modelling studies are commonly based on historical measured streamflow data, adjusted to represent flow conditions at pertinent locations for a specified condition of river basin development. The historical streamflow measurement needs to be adjusted to reflect conditions of river basin development at a specified past, present and future point in time. Some reservoir system modelling and analysis approaches are based on representation of the inflows. Based on the operation rules of water distribution units (WDU)/*tektek* in primary, secondary, tertiary, or quarternary canals, which have no operational gate deterministic or probabilistic models represent the streamflow analysis of each weir in Yeh Ho River. To estimate water availability, it will be necessary to determine a objective function on how the streamflow will be presented, which will largely determine the results and conclusions (Monzon *et al.*, 2015).

A deterministic model requires that inflows be simplified to an average, or a characteristic flow for each season, or sub-period of a representative year. This analysis was shown in the Figures 4.3 until 4.9. The result has been input to the simulation process by using the RIBASIM model.

7.1.3   Potential evapotranspiration

Evapotranspiration from crops and the soil surface is an important parameter in the water balance equation. It is dependent of climatic conditions, crop variety and stage of growth, soil moisture depletion and various physical and chemical properties of the soil. A two-measure procedure is generally followed in estimating evapotranspiration: (i) computation of the seasonal distribution of potential evapotranspiration (E_{tp}); (ii) adjustment of E_{tp} for crop variety and stage of growth (Walker and Skogerboe, 1987). A common practice for estimating evapotranspiration is to first estimate the reference evapotranspiration (E_{To}), and then to apply a crop coefficient (kc) (Trajkovic and Gocic, 2010).

The general features of evapotranspiration in the humid tropics as well as methods for the estimation of potential evapotranspiration (E_{tp}) were presented by Bruin (1983). In 1998 the FAO Irrigation and Drainage Paper No 56 *Crop Evapotranspiration* was published to revise the guidelines for computing crop water requirements (Allen *et al.*, 1998; Pereira *et al.*, 2015). Accordingly, Indonesia in the humid tropical region, has adopted this paper to determine the reference evapotranspiration, using Penman - Monteith in the Standard National Indonesia (RSNI T-01-2004). The equation and the results of this method are shown in Annex E.

7.1.4   Reservoir water surface losses and gains

Evaporation from the reservoir in each period (t) is based on the evaporation rate (er) with an average surface area of the water in the reservoir and the interval time, and is determined by the storage function (Wurbs, 1996):

$$E_i = A_i\ er \qquad\qquad\qquad 7\text{-}2$$

$$A_i = (A_t + A_{t + \Delta t})/2 \qquad\qquad\qquad 7\text{-}3$$

where:

E_i = evaporation in the reservoir each time to i

A_i = average surface area of the reservoir in time interval i

A_t values are specified in Δt and are a function of the volume of storage at the beginning and end of a time interval.

Rainfall and evaporation rates are often combined as a net rate. By way of ordinary unregulated streamflows provided as input to a reservoir system analysis model the corresponding net rainfall minus evaporation rates should reflect precipitation, which is not already accounted for in natural unregulated streamflows. With the reservoir, altogether the precipitation falling on the reservoir water surface is inflow. Net rainfall minus evaporation rates are sometimes adjusted to reflect the difference between rainfall falling on the reservoir water surface and runoff from rain falling on the land area at the site that contributes to streamflow before the reservoir was constructed. In this case the evaporation data are from Tiyinggading Station with a range of 0 - 13.5 mm/day.

7.1.5   Other reservoir losses and gains

Losses and gains of water to and from the ground under a reservoir are difficult to quantify. Losses to infiltration or seepage and gains from groundwater, or bank storage are typically considered negligible and ignored in reservoir system analysis studies (Wurbs, 1996). Most reservoirs are constructed at relatively impermeable locations. Permeability of the reservoir bottom is likely to decrease over time with sedimentation. The sediment deposits help cover the bottom and prevent seepage. Therefore in the present study such losses and gains have not been taken into account.

7.1.6   Reservoir elevation/storage/area relationship

In reservoir operation, the correlation between elevation, storage and area needed to control the active volume of reservoir storage, is obtained from reservoir data. It is defined by McMahon (2007) as difference between total storage capacity at full supply level and dead storage (the volume of water held below the lowest off-take). Telaga Tunjung Reservoir has an area-storage curve as shown in Figure 7.2.

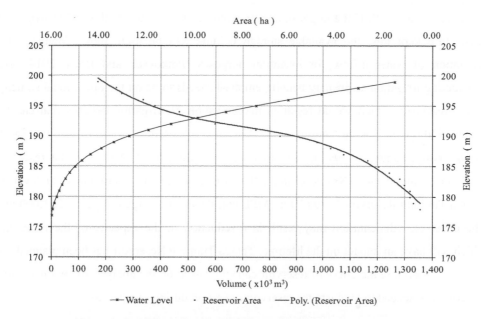

Figure 7.2. Reservoir elevation, storage and area relationship
(Regional River Office of Bali-Penida, 2006)

7.1.7   Flow routing in the reservoir and hydraulic profile of outlets

The operation of the hydraulic structures in Telaga Tunjung Reservoir has been evaluated and combined with the operation of the Subak irrigation schemes. The reservoir has one direct outlet to supply Subak Meliling and one direct outlet to supply domestic water of which the hydraulic profiles are shown in Annex F. The dimensions of the reservoir and the hydraulic structures are shown in Annex G. These dimensions are important in relation to the operation. The downstream Subak irrigation schemes receive released water through diversion weirs.

7.1.8   Evaluation of reservoir lifetime based on sedimentation

The sedimentation in the reservoir has been calculated to predict the lifetime. The results are presented in Annex H. The problem of sedimentation has to be challenged on the level of the reservoir itself, if the target minimum water levels reached at the end of irrigation season are kept low, the available storage volume will be high, thus allowing good regulation of water inflow for irrigation purposes (Petkovsek and Roca, 2014). By considering irrigation as primary benefit, emphasis needs to be on the operational strategy of the reservoir, which would result in maximization of irrigation benefits that can be achieved through storage conservation. Also for successful reservoir operation with storage conservation, an appropriate minimum operating level (MOL) of the reservoir needs to be maintained (Rashid *et al.*, 2014). Elevation of the minimum water level of Telaga Tunjung Reservoir is 190.70 m+MSL and the dead storage (50 years) 261,000 m^3. The onset of sedimentation may result in a reduction in the capacity of the reservoir, which will have an impact on the lifetime. Telaga Tunjung Reservoir has been planned for 50 years. Therefore the curve of reservoir capacity based on the analysis of sedimentation patterns was needed to evaluate the reservoir operation during this period.

7.2   Advanced irrigation node in RIBASIM

With the advanced irrigation node in RIBASIM it is possible to analyse and quantify Subak irrigation schemes with varying layers of paddy terraces, because the irrigated area, water demand, allocation, production costs and crop yield can be computed based on:

- cropping plan and crop characteristics;
- hydrological input;
- soil characteristics;
- topography and lay out of the irrigated area;
- operation and irrigation water management;
- production costs and crop yield;
- actual field water balance.

For the analysis it is useful and practical to make a distinction between the water delivery system: the river providing water to the Subak irrigation schemes, and the schemes with their water conveyance and distribution systems (Figure 7.3). Such a distinction corresponds with the typical management responsibility division between Government and farmers, or farmer's organizations.

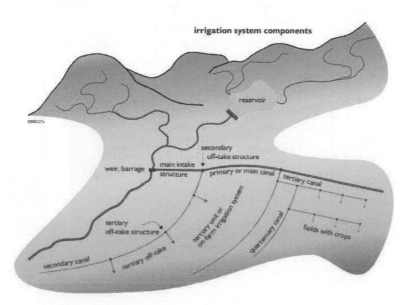

Figure 7.3. River basin and irrigation system components (Van der Krogt, 2008)

7.2.1   Schematization of the irrigated area

The advanced irrigation node represents a whole irrigated area or a part of an irrigated area, which in this study is a Subak irrigation scheme and Subak sub-schemes (water users association/*tempek*). The area is cultivated by a number of crops covering a certain percentage of the irrigated area. Further, the cultivated area of a particular crop is split into a number of sub-units (strips) (Figure 7.4). The width of a sub-unit depends on the progress of the land preparation and planting, and is computed for flood basin cultivations using the 'Van der Goor' method (Annex I).

The sub-unit can have a layered structure of fields for flood basin cultivation, allowing a flow of water from field to field. Figure 7.5 shows in a schematised way the field to field layered water distribution within the irrigated area.

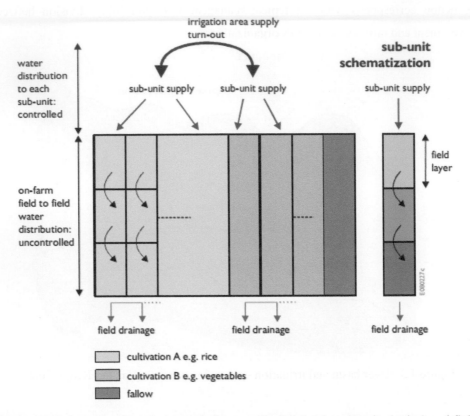

Figure 7.4. Schematization of an irrigated area: sub-division in sub-units (strips) and fields
(Van der Krogt, 2008)

The number of field layers can be specified in the input data. This indicates the extent, or effectiveness of the distribution system. If the number of layers is one then a perfect distribution system is assumed to be available. A large number of layers may indicate a complete lack of a distribution system and thus a full dependence on a field to field water distribution.

The water distribution within a Subak irrigation scheme can be represented more accurately with these layers allowing, for example, to represent the effect of canal systems and associated rotations and to estimate realistically the effect of water shortage. The modelling approach for the irrigation node is then the specification of the cropping plan (crop type, cultivated area, starting date, percolation pre-saturation requirements) and the number of layered fields for the flood basin crop, computation of a moisture balance for each field and water flows from field to field for the set of layered fields. A time staggering is adopted for crop cultivation in the set of layered fields.

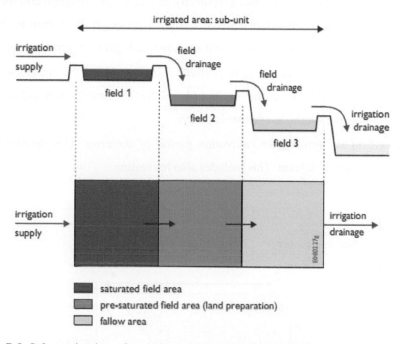

Figure 7.5. Schematization of an irrigated area sub-unit: field to field water flow for flood basin irrigated crops (Van der Krogt, 2008)

7.2.2   Interactive graphical cropping plan editor

The cropping plan is one of the input data requirements. The cropping plan consists of a table with the actual cultivation characterised by:

- cultivated crop;
- cultivated area (ha);
- starting date of the land preparation and transplanting period;
- percolation (mm/day);
- pre-saturation (mm).

In the model the cropping plan is defined for one year and is kept each year the same. At the determination of the cropping plan various constraints, such as crop characteristics, maintenance period(s), and overall water availability are taken into account. The cropping plan is presented graphically in a crop-time diagram and the water demand and availability are shown in a water balance hydrograph. Each crop in the crop time diagram is represented as a parallelogram as shown in Figure 7.6, characterized by:

- *the length of the land preparation period.* This determines the steepness of the parallelogram. At the start of the planting period the first farmers start and at the end of the planting period the last farmer is ready;

- *the length of the growing or cultivation period of the crop.* This determines the length of the parallelogram. This includes also harvesting.

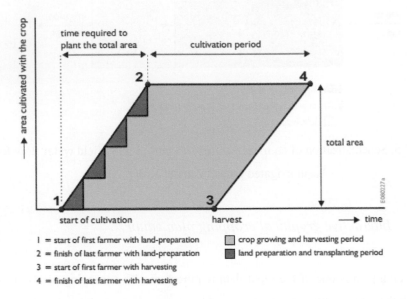

Figure 7.6. Format for presentation of a planned cultivation for a particular area
(Van der Krogt, 2008)

A combination of cultivations is scheduled in a cropping plan of the irrigated area. Figure 7.7 shows an example of the Subak Meliling cropping plan consisting of two cultivations. The crop is paddy by two varieties, which were determined based on the highest value of feasibility of farming (B/C ratio). These paddy varieties have the same land preparation and water requirement. Each cultivated area is split into a number of sub-units (strips) depending on the length of the planting period. RIBASIM contains an interactive graphical cropping plan editor for the composition of a feasible cropping plan at which water demand and availability are compared by a water balance hydrograph.

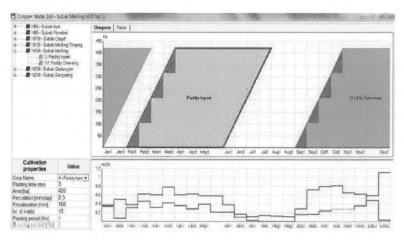

Figure 7.7. Schematization of the cropping plan and the water balance of Subak Meliling irrigation scheme with paddy crops

7.2.3   Simulation of a cropping plan

During the simulation and optimisation of Yeh Ho River, the cropping plans have been simulated by five scenarios as will be discussed in the next sections. Based on these scenarios the optimal irrigation performance has been determined as follows:

1.   with the RIBASIM model the water demand of the Subak irrigation systems based on six weirs along Yeh Ho River, the Telaga Tunjung Reservoir and one weir along the tributary Yeh Mawa River has been determined for each time step. The water availability at the weirs was based on the available water for the Subak irrigation schemes at 80% dependable flow over the period 2002 - 2010. The initial soil water content, the cropping plan, the survival fraction of the cultivated area, dependable rainfall, target level of the water layer or soil moisture, percolation, rainfall

effectiveness, irrigation practise and irrigation efficiency are taken into account;

2.   with the RIBASIM model the actual irrigation supplies for each irrigation node for the present time step, taking into account the water demand, the available water, the operation of the Telaga Tunjung Reservoir and the weirs, recoverable flow of the upstream schemes and the water allocation priority for each are computed;

3.   in the model, on a daily basis the actual soil water balances, based on the irrigation supplies, actual rainfall, actual crop evapotranspiration, percolation, field storage, seepage and drainage discharge are updated. The soil water content at the end of the time step becomes the new initial soil water content for the next time step. In this way, excess or shortage of irrigation water will be corrected in the next time step;

4.   finally, the crop survival fraction of each field, sub-unit, and cultivation based on the actual crop evapotranspiration are computed with the model. At the end of the growth period, the crop yield and production costs are computed.

7.2.4   Soil moisture characteristics

As stated before Bali has volcanic soils and Latosol soil is one of the two types of volcanic soil that cover the soil layer of the study region at Tabanan Regency (Sunarta, 2016). Latosol soil has the following characteristics (Saptaningsih, 2015; Yulia, 2015):

- the layer of soil is 1.3 to 5 metres thick, while the horizon is unclear;
- the colour of the soil is red, brown to yellowish. By the colour the natural fertility can be determined. More red is usually poorer. In general the nutrient content is from low to moderate;
- the soil is usually clay, while the organic content of the soil ranges from 3 - 9%. It is usually estimated at 5%;
- soil reaction is quite acid until acid between pH 4.5 - 6.5;
- the consistency of the grain structure is friable;
- infiltration and percolation are quite fast until quite slow;
- the water holding capacity is quite good;
- the soil is fairly resistant to the erosion.

Soils have physical characteristics such as texture, structure, bulk and particle density, porosity and permeability. Soils hold different amounts of water depending on their texture and structure. The most important physical characteristic is the field capacity (FC), especially in the root zone. This is the upper limit of the water holding capacity, while the lower limit is the permanent wilting point (PWP) (Zotarelli and Dukes, 2010).

In the paddy terraces of Meliling Subak irrigation scheme, soil samples have been taken and laboratory tests were done to describe the physical characteristics of the Latosol. The results of the analyses show that the soil texture is categorized by silty clay with medium until high plasticity and brown to yellowish colours. Moreover, the average value of the bulk density is 0.91 gr/m^3, particle density 2.58 gr/m^3, and porosity 0.65.

Based on the criteria of the Soil Laboratory of the University of Brawijaya the bulk density is categorized in the middle range (Bokings *et al.*, 2013). The bulk density is influenced by a solid soil, porous soil structure, texture, the availability of organic matter and land preparation, so that it quickly can be changed as a result of land preparation and the practice of cultivation (Hardjowigeno, 2003). The value of the porosity of Latosol soil is good with relatively limited pores to drain (Bokings *et al.*, 2013).

The soil moisture characteristics of Latosol and some other vocanic soils are shown in Figure 7.8. Latosol has a water content in the middle range, between Regosol as the lowest and Andosol as the highest at all pF values.

Andosol is categorized as the soil with a relatively high organic matter content. In contrast, Regosol has the most rough texture, and the water content is lower at all pF values. This shows that the soil moisture availability is influenced by the organic matter content and the soil texture. If the organic matter content of the soil is higher, then the soil moisture is higher, and the soil texture is more coarse, so the soil moisture is lower. The rate of decrease in the moisture content of the four soil types is shown in Figure 7.9. Regosol has the highest rate of decrease, followed by Podsolik MK, Latosol and Andosol.

In addition, tests with the double ring infiltrometer were conducted at four terraces in the early fallow season of July 2012. The tests have been conducted to investigate the infiltration rate or the permeability of the root zone ($\pm$ 20 cm). The permeability is used as an indicator for the drainage capacity of the soils. The average result of the root zone permeability of the paddy terraces was 75 cm/hour. This result is considered high

(Bokings *et al.*, 2013), which implies that the soil can rapidly drain excess water. This result shows that when the Subak farmers work for saturation before land preparation, the top soil of the paddy terraces need a high amount of water. This result can support the information on the irrigation practice of the Subak farmers related to land preparation.

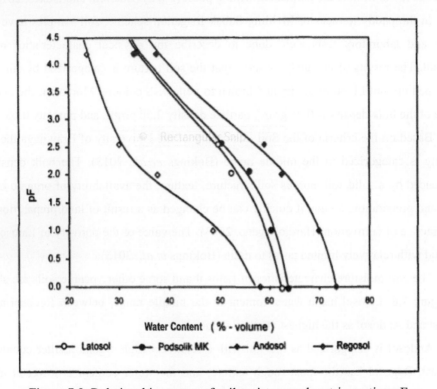

Figure 7.8. Relationship curves of soil moisture and matric suction pF

(Baskoro and Tarigan, 2007)

The type of soils with a low permeability is appropriate for rice fields. This implies that the hydraulic conductivity in saturated circumstances at a rice field should be quite low to prevent water losses, but still enough to leach the toxic materials (Emerson and Foster, 1985). The laboratory tests by the falling head method for soil depths of 2 up to 6 m resulted in an average hydraulic conductivity of 9.7 x 10^{-4} cm/hour or 2.7 x 10^{-9} m/s. This value represents the silty clay of the rice fields at the study location. The results of soil tests are shown in Table 7.1.

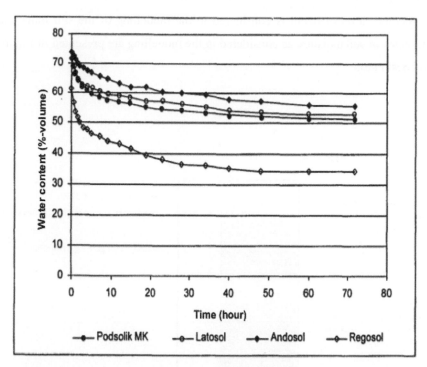

Figure 7.9. Moisture content under free drainage, related to the time after saturation
(Baskoro and Tarigan, 2007)

Table 7.1. Index soil properties of Latosol silty clay at the paddy terraces block

Depth (m)	2 - 3 m		4 - 5 m		5 - 6 m		Average
Soil index	Sample 1	Sample 2	Sample 1	Sample 2	Sample 1	Sample 2	
γ (kg/m^3)	1.64	1.64	1.53	1.53	1.51	1.51	1.56
ω (%)	0.71	0.72	0.61	0.62	0.81	0.84	0.72
Gs	2.63	2.67	2.41	2.45	2.64	2.64	2.58
γd (kg/m^3)	0.96	0.95	0.95	0.95	0.83	0.82	0.91
N	0.63	0.64	0.61	0.61	0.68	0.69	0.65
Sr	1.07	1.07	0.96	0.95	0.98	1.00	1.00
γsat (kg/m^3)	1.60	1.60	1.56	1.56	1.52	1.51	1.56
ωsat (%)	0.66	0.67	0.64	0.65	0.82	0.84	0.71
kv (cm/hour)	0.00091		0.0013		0.0007		9.7×10^{-4}

A medium level of detail is considered in the modelling of the soil moisture. The characteristics of soil moisture as considered in the modelling are presented in Figure 7.10 for flood basin crops.

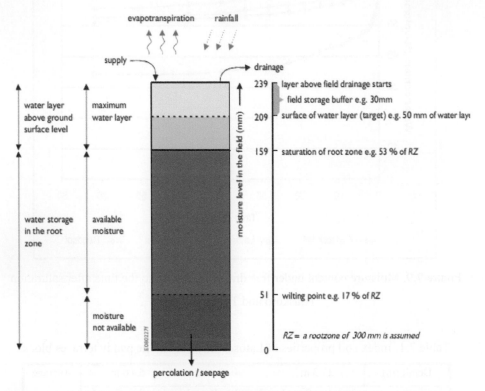

Figure 7.10. Schematization of water management at field scale for flood basin crops (Van der Krogt 2008)

7.2.5   Crop water requirement in the paddy terraces block

The computed consumptive use of paddy fields based on optimal soil moisture levels and the various factors depleting or replenishing soil moisture, is based on a set of fixed or pre-set conditions. Water requirements for a flood basin crop start with land preparation, which requires a large amount of water and constitutes peak flows in the canal system, and thus strongly determines design capacities (Van der Krogt, 2013).

As said, the RIBASIM model uses the 'Van der Goor' method to represent the increase in cultivated area during land preparation based on the Irrigation Planning Criteria (KP) 01 and the Technical Requirement (PT) 01 of the Department of Public Works. By this method, the non-linear increase in the area is associated with a constant supply of water during this period (Figures 7.11 and 7.12).

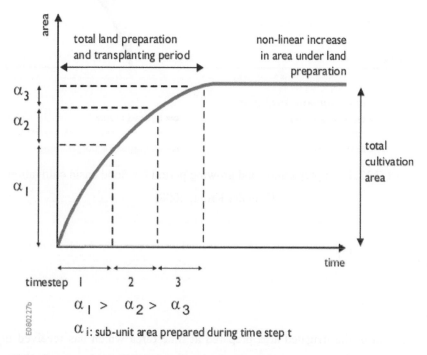

Figure 7.11. Progression of land preparation for flood basin crops (Van der Krogt, 2008)

In the land preparation period, the water requirements consist of two parts: (i) application of the pre-saturation and water layer requirements on the piece of land, which has been prepared during time dt; (ii) the evaporation and percolation over a progressively prepared piece of land. Water input to an irrigated area is kept constant during land preparation and results in a progression in land preparation according to equation 7-4. The water requirement for the land preparation and transplanting time step is shown in equation 7-5.

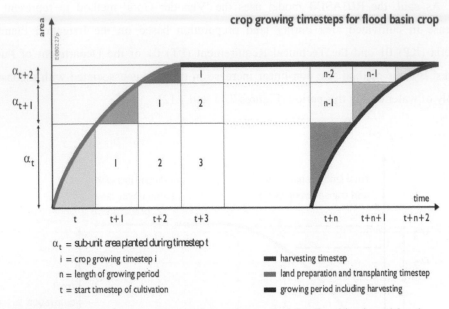

Figure 7.12. Land preparation and growing period for flood basin cultivation
(Van der Krogt, 2008)

$$Y = \left[\frac{e^{MT/S}}{(e^{MT/s} - 1)} \right] \cdot \left(1 - e^{-tM/S} \right)$$

7-4

where:

Y   = fraction of the irrigated area prepared at time (area which has received the pre-saturation, water layer and compensation for seepage and evaporation)

t   = time (days)

S   = water requirements for the pre-saturation of the field and water layer (mm)

M   = supply required for maintaining the water layer after pre-saturation is completed. This equals to the sum of evapotranspiration and percolation (mm/day)

T   = duration of pre-saturation (days)

e   = base of natural logarithm

$$\text{Req}_{\text{tot}} = S.Y' + (Cp.Eto + S).Y_{ep}'$$ 7-5

where:

Req_{tot} = total crop water requirement (mm/time step)

S = water requirement for the pre-saturation of the field and water layer (mm)

Y' = area prepared at the end of land preparation and transplanting time step

Cp = crop factor

Et_o = reference crop evapotranspiration

S = percolation

Y_{ep}' = factor indicating the cumulative area (evaporation and percolation)

The evaporation and percolation of a particular part of the irrigated area (saturated between t_0 and t) can be expressed as follows:

$$Y_{ep} = e^{MT/S}.\left(e^{MT/S} - 1\right) x \left[e^{-t_0.M/S}.\left(t - t_0 - S/M\right) + S/M.e^{-M.t/S} \right]$$ 7-6

where:

Y_{ep} = proportion of pre-saturated area

t_0 = starting time of the period between t_0 and t for which pre-saturation must be completed

t = time (days)

S = water requirement for the pre-saturation of the field and water layer (mm)

M = supply required for maintaining the water layer after pre-saturation is completed. This equals to the sum of evapotranspiration and percolation (mm/day)

T = length of pre-saturation period (days)

e = base of natural logarithm

A = area under the curve with the proportion of the pre-saturated area versus time

7.2.6 Computation of command area water demand, actual field water balance and effective irrigation water supply

The net field water requirements are obtained by subtracting the effective contribution of rainfall, or from the computed field water requirement of the crops. An efficiency factor is then applied to the total net field water requirement in order to find the water demand at the intake. The effectiveness of rainfall depends on the water supply to the irrigated area and the actual moisture condition in the fields.

The efficiency of water supply is a variable depending on the source and type of irrigation. The efficiency increases under dry conditions. It is for example known that the efficiency may go up to almost 100% during periods of shortage. For a flood basin the approach followed in the DelftAGRI model is schematically shown in Figure 7.13. DelftAGRI accounts for irrigation requirements based on farming and irrigation practices, physical parameters related to soils and hydro-meteorological characteristics, crop damage and production costs (Borden *et al.*, 2016).

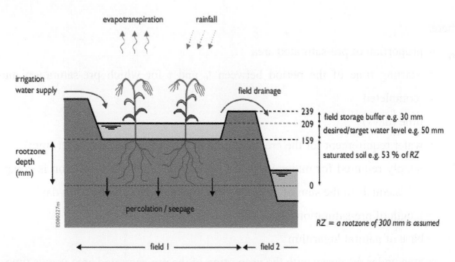

Figure 7.13. Schematization of the field water balance for flood basin crops

(Van der Krogt, 2008)

An average application efficiency of 80% has been considered and a maximum efficiency during drought conditions of 90%. In the determination of the irrigation water supplies, an irrigation efficiency of 80% has been applied to the net field water requirements for all the crops. For the primary and secondary canals, an efficiency of 80 - 90% has been applied. This means that it is assumed that 80 - 90% is effectively available to the plants in the soil and 10 - 20% will be lost to evaporation and seepage in canals and in the distribution system. In the Subak irrigation schemes, most of the confluence flows are located after the primary and secondary canals and one confluence system in the upstream of Yeh Ho River as shown in Figure 7.14 (Subak Penebel to Subak Benana). As a result the tertiary, quarternary and lower level canals almost only receive recoverable flow from the upper Subak irrigation schemes, so the water losses are more or less in balance or higher than the supply from the weirs. Based on Chapter 5, the observed outlets which also function to provide recoverable flow have a maximum of 83%, an average of 62% and a minimum of 21% from the observed inlets. The results of the observations during two dry and two wet seasons are shown in Annex J.

Figure 7.14. Confluence flow as recoverable flow supports downstream irrigation schemes

In the performance simulation of the cropping plan over a certain period the net command area water demand has been computed by the water requirements of each field and by simulating the cascade flow from field to field. For the computation of the demand only the actual cultivated area (survival fraction of potential cultivated area) is taken into account. RIBASIM provides also an option to simulate the level of irrigation system management: with and without feedback from actual field conditions. Operation with feedback has the following effect on the water demand:

- *actual rainfall > dependable rainfall:* the actual water level in the field above the target level (storage of water in the soil) which means that in the next time step the demand will be lower;

- *actual rainfall < dependable rainfall:* the actual water level in the field drops below the target level which means that in the next time step the demand will be higher.

The computation of the actual field water balance has been carried out for each field on a daily basis over the time step. This was based on the initial soil moisture content, actual water supply, actual rainfall, evapotranspiration and percolation or seepage. Drainage has been computed as a rest term: if the storage in the field, after considering the above components, is above the desired level and the field storage buffer then the extra water is considered as drainage. If the root zone soil moisture drops below the wilting point then actual evapotranspiration and percolation are considered to be zero.

Actual irrigation water supply and rainfall enter both as input into the computation of the water balance. If the sum of the irrigation water supply and rainfall is larger than the requirement then there is in fact no way of telling which input contributes how much to drainage. The following approach has been taken to separate the contributions of irrigation water supply and rainfall to the drainage, or in other words to determine the effective contribution of rainfall and irrigation water supply to the consumptive demand (evapotranspiration and percolation or seepage).

If the field water level drops below the target level for a certain lower field then the field storage of the upper fields is levelled over all fields. Drainage from the field will occur if for flood basin crops the field water level reaches the field storage level or if for dryland crops the soil moisture reaches the soil moisture storage. The outflow goes to the next lower field. The drainage from the lowest field is added to the drainage flow out of the irrigation node.

7.3   Results of economic evaluation of storage allocation

7.3.1 Pricing of paddy productivity

The Subak irrigation schemes are suitable for a non-volumetric method of irrigation water pricing. The non-volumetric methods charge for irrigation water is based on a per output basis, a per-input basis, a per-area basis, or on land values. These methods are easy to implement and to administer and are best suited to continuous flow irrigation, which may explain the prevalence in the irrigation world (Easter and Welsch, 1986; Easter and Tsur, 1995; Johansson, 2000). The market based method mechanisms have recently arisen as a need to address water pricing inefficiencies inherent in existing irrigation institutions. Subak Associations only charge the operation and maintenance cost of the tertiary, quarternary and lowest level of the system, which need to be repaired incidentally. However, it is related to the value of water for irrigation of paddy terraces. Irrigation water is not charged for the transfer process.

In the Dutch colonial period, at April 25, 1939, a food agency, called Voedings Middelen Fonds (VMF), was established, which aimed to control fluctuations of the price of rice that were quite sharp in 1919/1920, and that had declined harshly in 1930, so that the farmers found it difficult to pay taxes. In May 10, 1967 a government agency, the Bureau of Logistics (*Badan Urusan Logistik* (BULOG)), was established to control prices in the rice market (*Harga Pokok Penjualan*), During the Suharto era *(Orde Baru)* (30 years), the success of BULOG was characterized by centralization of management with strong leadership for implementing the policy. However, due to Indonesia's economic policy shift towards openness, deregulation of international trade, banking and finance, the closed and centralized management system of BULOG started to loose effectiveness and confidence. Since the early 1990s, the performance of BULOG has attracted criticism from economists and other social scientists (Rachman and Purwoto, 2005).

Since 2003, the status of BULOG has changed from a government agency into a Public Company (*Perusahaan Umum (PERUM)*), and it does not have sole authority to implement pricing policies anymore (Asnawi, 2015). Since then most of the Subak farmers have sold their products directly to middleman, who can pay a higher price than BULOG. On the other hand, some Subak farmers have rice mills themselves and sell directly to the market.

The simulation has been based on the yield and production cost per node as mentioned for a Subak irrigation scheme. The analysis of the primary data was based on the period of two wet seasons and two dry seasons in two Subak irrigation schemes: Subak Caguh and Subak Meliling. The results are shown in Table 7.2. RIBASIM computes an agro-economic model with cost figures for the situation without a significant reduction of yields due to drought. Damage can be considered as potential production costs. In case of major crop losses due to water shortages, it was expected that production costs will also be reduced.

7.3.2   Pricing of domestic water

The pricing of domestic water is determined by the Regency Government under the Local Water Supply Utility (*Perusahaan Daerah Air Minum (PDAM)*) as regional owned corporation (*Badan Usaha Milik Daerah (BUMD)*). In this research, the pricing of domestic water from Telaga Tunjung Reservoir was input, because the domestic water is processed in the water treatment plant (WTP) under supervision of PDAM Tabanan Regency at the reservoir location. The pricing of domestic water basic cost has been based on the annual operation and maintenance of the reservoir. It was Rp. 564 per m^3 in 2013 and Rp. 941 per m^3 in 2014. The PDAM Tabanan Regency charged a constant price of Rp. 1,300 per m^3 during the recent years.

Table 7.2. Yield and production cost analysis in Indonesian Rupiah

Subak scheme	Subak Caguh							Subak Meliling			
Observation location	Tempek Jangkahan 2013 and 2015			Angligan 2014				Angligan 2015			
Area (ha)	0.8	0.8	0.8	0.6	0.37	1.3	0.3	1.5	0.45	0.2	1.35
Start land preparation	16-5-13	5-12-13	1-2-15	24-12-14	26-12-14	27-12-14	29-12-14	16-2-15	4-2-15	02-2-15	04-2-15
Harvest time	13-9-13	19-3-14	21-5-15	12-4-15	13-4-15	20-4-15	12-4-15	24-5-15	17-4-15	26-4-15	07-4-15
Paddy Crop	Ciherang Prima	Ciherang Prima	Inpari 19	Waengapu	Ciherang Prima	Ciherang Prima	Inpari 19	Serang	Serang 64	Sumatera 19	Inpari 19
Unhusked harvest (kg)	4,560	4,630	3,000	1,400	1,300	3,900	1,200	5,500	1,110	600	7,000
Revenue of harvest (Rp.)	16,200	17,600	9,000	9,600	5,180	13,300	4,800	28,900	3,770	2,580	27,300
Price per kg	3.55	3.79	3	6.86	4	3.40	4	5.25	3.40	4.30	3.90
kg/ha	5,700	5,790	3,750	2,330	3,500	3,000	4,000	3,670	2,470	3,000	5,190
Revenue per ha (Rp.)	20,200	22,000	11,300	16,000	14,000	10,200	16,000	19,300	8,390	12,900	20,200
Cost (Rp.)											
a. ceremonies	300	250	300	200	185	200	200	2,300	200	200	300
b. land preparation	1,160	1,450	1,840	1,200	740	2,600	600	4,400	1,500	400	600
c. planting		640	800								
d. seeding	1,200	150	1,030	350	355	1,330	400	1,680	690	345	500
e. weeding	440	600	630	860	790	1,500	301	350		335	235
f. spraying	955	720	770	525	600	1,000		1,500		200	140
Total cost (Rp.)	4,050	3,810	5,370	3,140	2,670	6,630	1,500	10,200	2,390	1,480	1,780

Reservoir operation for water supply to Subak irrigation schemes in Yeh Ho River Basin

Table 7.2. *continued*

	0.89	0.82	1.79	2.24	2.06	1.70	1.25	1.86	2.15	2.47	0.25
Cost per kg (Rp.)	0.89	0.82	1.79	2.24	2.06	1.70	1.25	1.86	2.15	2.47	0.25
Cost per ha (Rp.)	5,060	4,760	6,710	5,230	7,220	5,100	5,000	6,820	5,310	7,400	1,320
Benefit per ha (Rp.)	15,200	17,200	4,540	10,800	6,780	5,100	11,000	12,500	3,080	5,500	18,900
Benefit per kg (Rp.)	2.7	3.0	1.2	4.6	1.9	1.7	2.8	3.4	1.3	1.8	3.7
R/C (Efficiency of farming)	4.0	4.6	1.7	3.1	1.9	2.0	3.2	2.8	1.6	1.7	15.4
B/C (Feasibility of farming)	3.0	3.6	0.7	2.1	0.9	1.0	2.2	1.8	0.6	0.7	14.4
Rupiah in thousand	1 US $ = Rp. 13,000			standard cost = 500 US$/ha							

7.4 Scenario analysis, simulation and optimisation of Yeh Ho River Basin

The scenarios for the Subak schemes within Yeh Ho River Basin are based on the flowchart as shown in Figure 6.1 and the cropping pattern in Table 4.1. The table shows the total functional paddy field area that has been used in the model simulation, although the data of the Department of Agriculture and Horticulture, Tabanan Regency differ. The area of agriculture fields was 22,388 ha in 2013 and 21,962 ha in 2014, it decreased with 1.9%. In 2015, the agriculture fields were 21,742 ha, and decreased again with 1.0%.

While the start of the land preparation plays a crucial role in the water management the optimisation has focused on variations in the start of land preparation. The start of land preparation is based on the cropping pattern agreement in *Awig-awig* Subak as it has been compromised by the Subak farmers. In the reality at the field, there sometimes have been sudden changes in the start of land preparation, such as because of the hydro-climate and the maintenance of the canals in the concerned years, as it occurred in the case for two years period of the observations in this study (April 2013 - April 2015). Although the change of cropping pattern is usually prepared among the farmers and the Subak leader (*Pekaseh*), occasionally, individual Subak farmers start the land preparation by considering that their WDU (*tektek*) is having available water. However, the reason is the justice of the *THK* philosophy that is based on *Awig-awig* Subak related to irrigation water distribution, and mentions the farmer's right on his section of paddy field of approximately 0.5 - 1 ha

The concept of Subak water management is a boundary condition for the simulation process. With the simulation based on alternatives for scheduling of the starting time of land preparation in the paddy cultivation season (*nyorog)* the benefit of crop products, and the indigenous cropping patterns based on the blocks upstream (*ngulu*), midstream (*maongin*) and downstream (*ngasep*) has been optimised.

All scenarios were based on the 80% dependable discharge in the Yeh Ho River, as mentioned in Chapter 4. Following the approach for input in the simulation is the type of paddy crops, which have been chosen based on two highest values of feasibility of farming (B/C ratio): Inpari 19 and Ciherang Prima as shown in Table 7.3. Another approach for input in the simulation of RIBASIM is the percentage of recoverable flow from the upper to the lower of paddy terraces schemes. For all scenarios, the average 62% (normal condition) and minimum 21% (dry condition) for the recoverable flow have been applied in the simulation.

Table 7.3. Recapitulation of average yield and production costs

Paddy crops	kg/ha	Price/kg (Rp.)	Cost/ha (Rp)	R/C	B/C
Ciherang prima	4,500	3,700	5,360,000	3.1	2.1
64 Sumatera	3,000	4,300	7,400,000	1.7	0.7
Serang	3,070	4,350	6,060,000	2.2	1.2
Inpari 19	4,310	3,650	4,340,000	3.6	2.6
Waengapu	2,330	6,850	5,230,000	3.1	2.1

The RIBASIM model can simulate alternative numbers of terraces. Therefor the simulation process could also have been aimed on how the influence of the number of terraces for the recoverable flow from the upper schemes can support the lower schemes, especially within the schemes in the upstream of Yeh Ho River Basin: Aya, Penebel, Riang, Jegu and Caguh. The preliminary results did not show significant differences in the optimum of agriculture production and little differences between the scenarios from scheme to scheme. The differences in confluence flows between 25 and 35 terraces from scheme to scheme were 0.001 - 0.002 m³/s. While the objective of the study was optimisation of agriculture productivity all scenarios have been based on 25 terraces, which were described in the paddy terraces block in Chapter 5.

Finally, the scenario analysis aimed to determine a recommendation for the Subak farmers, how the available water based on the start of land preparation (*nyorog*) can give optimum agriculture production for all schemes of Subak Agung Yeh Ho.

7.4.1   Simulation of the first scenario

In the first scenario the land preparation starts in period 13 (July I) for Subak Aya. The cropping pattern is shown in Table 7.4, which was legally announced prior to the start of the operation of Telaga Tunjung Reservoir in 2006. The simulation of the first scenario shows the cropping pattern diagram in Subak Aya, Subak Penebel (one weir system with Subak Riang and Jegu), Subak Caguh and Subak Meliling-Timpag in the upstream of the Telaga Tunjung Reservoir. These Subak systems should be concerned, because of the allocation of water from the upstream to the midstream of Yeh Ho River has been insufficient for irrigation since a long time, when the utilization of Gembrong Spring in the upstream was mainly used for domestic water.

The water balance between weir supply and crop water requirement of Subak Aya (Figure 7.15) shows deficit irrigation during Jan I - Mar II and Jul II - Oct II. Subak Penebel (Figure 7.16) shows a shift trend that has deficit irrigation during Jan I - Apr I and Aug I - Nov II. Moreover, the water balance of Subak Caguh (Figure 7.17) has a longer and higher trend than the previous schemes, there was deficit irrigation during Jan I - May I and Aug I - Nov II. There was no problem with irrigation in Subak Meliling-Timpag, which is shown in Figure 7.18. However, that scheme only serves 142 ha of paddy field. In the upstream and midstream of Yeh Ho River, the largest scheme is Subak Caguh. Therefore the irrigation water distribution within this scheme is challenging.

The confluence flows from diverted flow and recoverable flow are considered to be 62 and 21% of recoverable flow for each scheme. The results show significant differences in confluence flows (Figure 7.19), which ranged from Subak Aya to Subak Penebel from 1.24 to 1.11 m^3/s, Subak Penebel and Subak Caguh have 1.33 and 1.08 m^3/s, Subak Caguh and Subak Meliling-Timpag have 0.97 and 0.71 m^3/s. Following the Telaga Tunjung Reservoir, Subak Meliling has two confluence flows: to Yeh Ho River of 0.05 and 0.01 m^3/s, and to Subak Sungsang 1.24 and 1.17 m^3/s. Furthermore, the results of the simulation of water allocation and agriculture production are explained in Table 7.5. The actual field level production has an average percentage of 92.0% for 62% and 91.5% for 21%.

Table 7.4. The cropping pattern of the first scenario in Subak Agung Yeh Ho

Subak irrigation schemes	Functional paddy fields	Blocks			When to start land preparation Paddy I and Paddy II (24 periods per year)
		Upstream (*Ngulu*)	Midstream (*Maongin*)	Downstream (*Ngasep*)	
	ha	Ha	ha	ha	
1. Aya	644	644			Jul I (13) and Dec I (23)
2. Penebel	731	731			Jul II (14) and Dec II (24)
3. Riang	25	25			Aug I (15) and Jan I (1)
4. Jegu	111	111			Aug II (16) and Jan II (2)
5. Caguh	1093		1093		Jan I (1) and Aug I (15)
6. Meliling-Timpag	142		142		Jan II (2) and Aug II (16)
7. Telaga Tunjung Reservoir:					
• Meliling	420		420		Feb I (3) and Sep I (17)
• Sungsang	430			430	Feb I (3) and Oct I (19)
• Gadungan	485		485		Feb II (4) and Sep II (18)

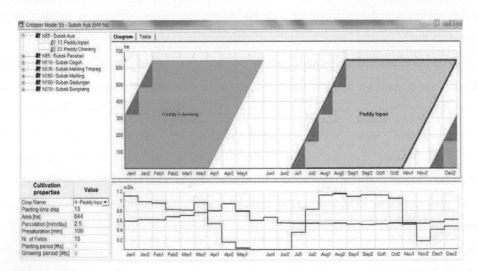

Figure 7.15. Cropping pattern and water balance of the first scenario for Subak Aya

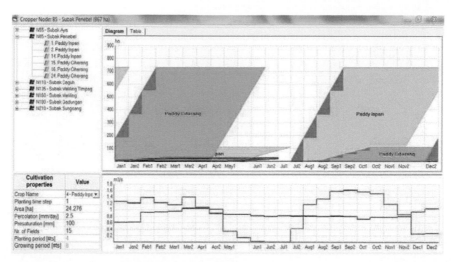

Figure 7.16. Cropping pattern and water balance of the first scenario for Subak Penebel

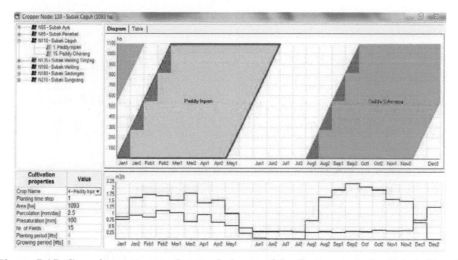

Figure 7.17. Cropping pattern and water balance of the first scenario for Subak Caguh

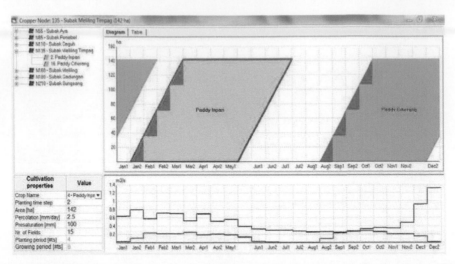

Figure 7.18. Cropping pattern and water balance of the first scenario for Subak Meliling-Timpag

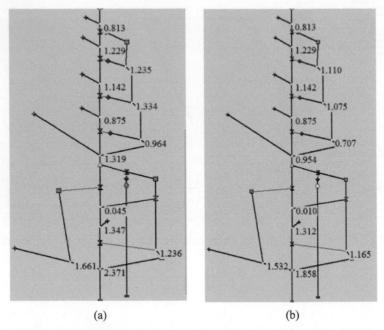

(a)   (b)

Figure 7.19. Confluence flows for the first scenario at 62 and 21%

Table 7.5. Water allocation and agriculture production of the first scenario

Node	Name Subak Scheme	Yearly average				Success time steps		Success years		Pot.field level prod.		Act.field level prod.			Act.prod. Costs
		Demand (MCM)	Shortage (m³/s)	Demand (MCM)	Shortage (m³/s)	number (-)	rate (%)	number (-)	rate (%)	Ton	10^6 Rp	Ton	10^6 Rp	(%)	10^6 Rp
62 % of recovarable flow															
1	Subak Aya	25.0	10.91	0.8	0.33	19	38.0	0	0	4,230	15,500	3,700	13,600	87.5	4,400
2	Subak Penebel	34.7	7.64	1.1	0.23	26	52.0	0	0	5,700	20,900	5,180	19,000	91.0	6,110
3	Subak Caguh	43.3	12.78	1.3	0.39	25	50.0	0	0	7,270	26,800	6,160	22,700	84.7	7,830
4	Subak Meliling Timpag	5.8	0.00	0.2	0.00	50	100.0	2	100	945	3,480	945	3,480	100.0	1,070
5	Subak Meliling	16.3	2.63	0.5	0.08	39	78.0	0	0	2,790	10,290	2,790	10,300	100.0	3,160
6	Subak Gadungan	18.7	2.25	0.6	0.07	42	84.0	0	0	3,230	11,900	3,230	11,900	100.0	3,650
7	Subak Sungsang	14.6	1.78	0.4	0.05	43	86.0	0	0	2,860	10,500	2,860	10,500	100.0	3,240
	Total	158.5	38.0	4.8	1.16					27,000	99,400	24,900	91,500	92.0	29,500
21 % of recovarable flow															
1	Subak Aya	25.0	10.91	0.8	0.33	19.0	38.0	0	0	4,230	15,500	3,700	13,600	87.5	4,400
2	Subak Penebel	34.7	11.17	1.1	0.34	24.0	48.0	0	0	5,700	20,900	5,130	18,800	90.0	6,080
3	Subak Caguh	42.9	21.54	1.3	0.66	18.0	36.0	0	0	7,270	26,800	6,060	22,300	83.3	7,750
4	Subak Meliling Timpag	5.8	0.00	0.2	0.00	50.0	100.0	2	100	945	3,480	945	3,480	100.0	1,070
5	Subak Meliling	16.1	5.51	0.5	0.17	28.0	56.0	0	0	2,790	10,290	2,800	10,300	100.0	3,160
6	Subak Gadungan	18.5	5.24	0.6	0.16	32.0	64.0	0	0	3,230	11,900	3,230	11,900	100.0	3,650
7	Subak Sungsang	14.9	3.85	0.5	0.12	39.0	78.0	0	0	2,860	10,500	2,860	10,500	100.0	3,240
	Total	158.0	58.2	4.8	1.77					27,000	99,400	24,700	90,900	91.5	29,400

7.4.2   Simulation of the second scenario

In the second scenario the land preparation starts in period 14 (July II) for Subak Aya and also a sequence change for the other schemes, with 15 days shift from the first scenario (Table 7.6). The Subak schemes before the Telaga Tunjung Reservoir remain to require attention for their cropping pattern and the water balance. The water balance between weir supply and crop water requirement of Subak Aya is shown in Figure 7.20. It has the same trend as in the first scenario with deficit irrigation during Jan I - Mar II and Aug I - Nov II. Otherwise, in this scenario, Subak Penebel has a different trend of deficit irrigation, while it shows a shift trend during Jan I - Apr II and Aug II - Nov II (Figure 7.21).

Table 7.6. The cropping pattern of the second scenario in Subak Agung Yeh Ho

Subak irrigation schemes	Functional paddy fields	Blocks			When to start land preparation Paddy I and Paddy II (24 periods per year)
		Upstream (*Ngulu*)	Midstream (*Maongin*)	Downstream (*Ngasep*)	
	ha	ha	ha	ha	
1. Aya	644	644			Jul II (14) and Dec II (24)
2. Penebel	731	731			Aug I (15) and Jan I (1)
3. Riang	25	25			Aug II (16) and Jan II (2)
4. Jegu	111	111			Sep I (17) and Feb I (3)
5. Caguh	1093		1093		Jan II (2) and Aug II (16)
6. Meliling-Timpag	142		142		Feb I (3) and Sep I (17)
7. Telaga Tunjung Reservoir:					
• Meliling	420		420		Feb II (4) and Sep II (18)
• Sungsang	430			430	Feb II (4) and Oct II (20)
• Gadungan	485		485		Mar I (5) and Oct I (19)

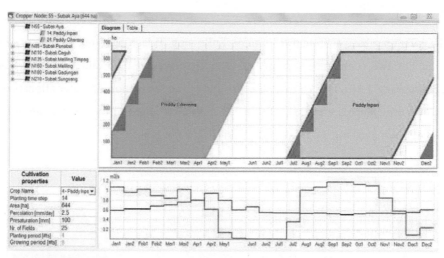

Figure 7.20. Cropping pattern and water balance of the second scenario for Subak Aya

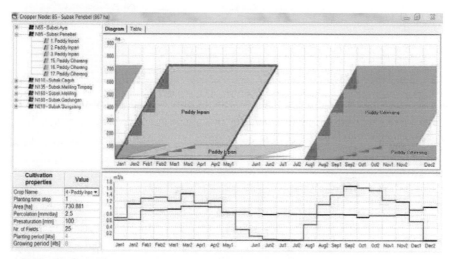

Figure 7.21. Cropping pattern and water balance of the second scenario for Subak Penebel

The water balance of Subak Caguh (Figure 7.22) shows a shift trend of deficit irrigation during Feb I - May II and Aug II - Dec I. At last, Subak Meliling-Timpag still has no deficit irrigation (Figure 7.23).

The principle of justice based on *THK* portrays how Subak farmers are concerned 'when and how' the irrigation water comes. The results show significant differences in confluence flows (Figure 7.24), from Subak Aya to Subak Penebel of 1.21 and 1.10 m³/s

Subak Penebel and Subak Caguh have 1.28 and 1.06 m³/s, Subak Caguh and Subak Meliling-Timpag have 0.95 and 0.71 m³/s. Following Telaga Tunjung Reservoir, Subak Meliling has two confluence systems: to Yeh Ho River of 0.05 and 0.01 m³/s, and to Subak Sungsang 1.34 and 1.26 m³/s. The results of the simulation of water allocation and agriculture production are given in Table 7.7 and the averages of the percentage of actual field level production are 94.3% for 62% and 93.8% for 21%.

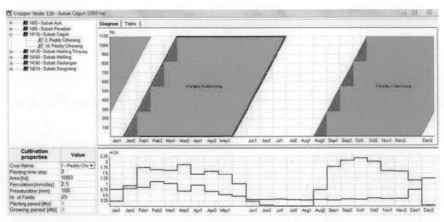

Figure 7.22. Cropping pattern and water balance of the second scenario for Subak Caguh

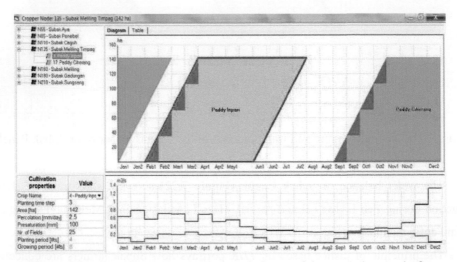

Figure 7.23. Cropping pattern and water balance of the second scenario for
Subak Meliling-Timpag

Table 7.7. Water allocation and agriculture production of the second scenario

Node	Name Subak Scheme	Yearly average				Success time steps		Success years		Pot. field level prod.		Act. field level prod.			Act. prod. costs
		Demand (MCM)	Shortage (m³/s)	Demand (MCM)	Shortage (m³/s)	numberrate (-)	(%)	numberrate (-)	(%)	Ton	10⁶ Rp	Ton	10⁶ Rp	(%)	10⁶ Rp
62 % of recovarable flow															
1	Subak Aya	25.9	11.41	0.8	0.35	18	36.0	0	0	4,230	15,500	3,730	13,700	88.2	4,420
2	Subak Penebel	35.5	7.37	1.1	0.23	27	54.0	0	0	5,760	21,200	5,380	19,800	93.4	6,520
3	Subak Caguh	42.7	12.97	1.3	0.40	29	58.0	0	0	7,370	27,300	6,700	24,800	90.9	8,730
4	Subak Meliling Timpag	5.7	0.00	0.2	0.00	50	100.0	2	100	945	3,480	945	3,480	100.0	1,070
5	Subak Meliling	16.7	2.34	0.5	0.07	38	76.0	0	0	2,790	10,300	2,790	10,300	100.0	3,160
6	Subak Gadungan	19.0	1.89	0.6	0.06	42	84.0	0	0	3,230	11,900	3,230	11,900	100.0	3,650
7	Subak Sungsang	16.9	1.85	0.5	0.06	43	86.0	0	0	2,860	10,500	2,860	10,500	100.0	3,240
	Total	162.3	37.8	4.9	1.15					27,200	100,000	25,600	94,500	94.3	30,800
21 % of recovarable flow															
1	Subak Aya	25.9	11.41	0.8	0.35	18	36.0	0	0	4,230	15,500	3,730	13,700	88.2	4,420
2	Subak Penebel	35.7	11.39	1.1	0.35	19	38.0	0	0	5,760	21,200	5,340	19,600	92.6	6,520
3	Subak Caguh	42.9	22.10	1.3	0.67	19	38.0	0	0	7,370	27,300	6,600	24,400	89.5	8,660
4	Subak Meliling Timpag	5.7	0.00	0.2	0.00	50	100.0	2	100	945	3,480	945	3,480	100.0	1,070
5	Subak Meliling	16.5	5.41	0.5	0.17	29	58.0	0	0	2,790	10,300	2,790	10,300	100.0	3,160
6	Subak Gadungan	19.1	4.69	0.6	0.14	33	66.0	0	0	3,230	11,900	3,230	11,900	100.0	3,650
7	Subak Sungsang	17.2	5.18	0.5	0.16	36	72.0	0	0	2,860	10,500	2,860	10,500	100.0	3,240
	Total	162.9	60.2	5.0	1.83					27,200	100,000	25,500	93,900	93.8	30,700

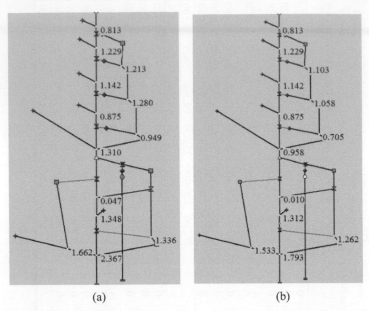

(a)   (b)

Figure 7.24. Confluence flows for the second scenario at 62 and 21%

7.4.3   Simulation of the third scenario

In the third scenario the land preparation starts in period 15 (Aug I) for Subak Aya and in sequence for the other schemes, with 15 days following from the second scenario. The cropping pattern is shown in Table 7.8. Subak schemes upstream of TelagaTunjung Reservoir still focus for their cropping pattern and the water balance between weir supply and crop water requirement. The water balance of Subak Aya (Figure 7.25) has the same trend as in the first and second scenarios, but in the third scenario there is a quite different trend with deficit irrigation during Jan II - Apr I and Aug II - Nov II.

In addition, Subak Penebel has a shifting trend compared to the two previous scenarios that shows deficit irrigation during Feb I - May I and Sep I - Nov II (Figure 7.26). The different trends occurred in Subak Caguh (Figure 7.27) which early in the year, Jan I, has deficit irrigation, but in Jan II there is no deficit, then from Feb II until June I, and during Sep I - Dec I there is deficit irrigation. At last, Subak Meliling-Timpag still has no deficit irrigation (Figure 7.28).

Table 7.8. The cropping pattern of the third scenario in Subak Agung Yeh Ho

Subak irrigation schemes	Functional paddy fields	Blocks			When to start land preparation Paddy I and Paddy II (24 periods per year)
		Upstream (*Ngulu*)	Midstream (*Maongin*)	Downstream (*Ngasep*)	
	ha	ha	Ha	ha	
1. Aya	644	644			Aug I (15) and Jan I (1)
2. Penebel	731	731			Aug II (16) and Jan II (2)
3. Riang	25	25			Sep I (17) and Feb I (3)
4. Jegu	111	111			Sep II (18) and Feb II (4)
5. Caguh	1093		1093		Feb I (3) and Sep I (17)
6. Meliling-Timpag	142		142		Feb II (4) and Sep II (18)
7. Telaga Tunjung Reservoir:					
• Meliling	420		420		Mar I (5) and Oct I (19)
• Sungsang	430			430	Mar I (5) and Nov I (21)
• Gadungan	485		485		Mar II (6) and Oct II (20)

In operation the Subak farmers of the irrigation systems also make use of the recoverable flow from the upstream schemes. By shifting the start of the land preparation, in the third scenario there is also a shift in the results. The results of confluence flows (Figure 7.29) are the same as with the second scenario, which is expected in Subak Aya to Subak Penebel to be 1.21 and 1.10 m^3/s, Subak Penebel to Subak Caguh 1.28 and 1.06 m^3/s and Subak Caguh to Subak Meliling-Timpag 0.95 and 0.71 m^3/s.

Moreover after the Telaga Tunjung Reservoir Subak Meliling has quite different results compared to the second scenario: to Yeh Ho River 0.05 and 0.01 m^3/s, and to Subak Sungsang 1.29 and 1.21 m^3/s. The results of the simulation of water allocation and agriculture production are explained in Table 7.9, that has an average actual field level production of 98.0% for 62% and 97.7% for 21%.

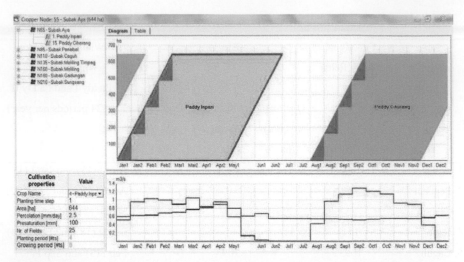

Figure 7.25. Cropping pattern and water balance of the third scenario for Subak Aya

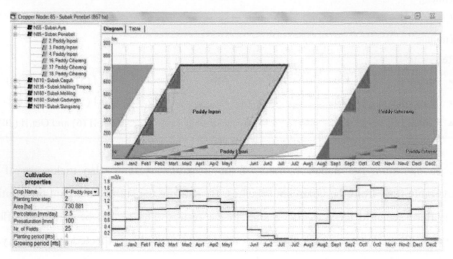

Figure 7.26. Cropping pattern and water balance of the third scenario for Subak Penebel
867 ha

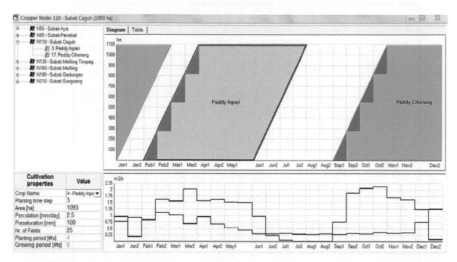

Figure 7.27. Cropping pattern and water balance of the third scenario for Subak Caguh

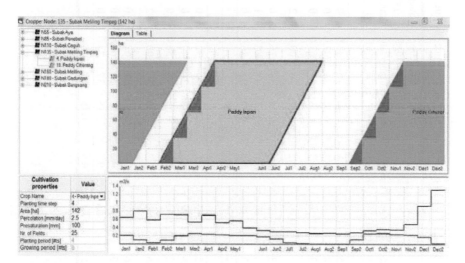

Figure 7.28. Cropping pattern and water balance of the third scenario for

Subak Meliling-Timpag

Table 7.9. Water allocation and agriculture production of the third scenario

Node	Name Subak Scheme	Yearly average				Success time steps		Success years		Pot. field level prod.		Act. field level prod.			Act. prod. costs
		Demand (MCM)	Shortage (m³/s)	Demand (MCM)	Shortage (m³/s)	number (-)	rate (%)	number (-)	rate (%)	Ton	10⁶ Rp	Ton	10⁶ Rp	(%)	10⁶ Rp
62 % of recoverable flow															
1	Subak Aya	26.5	11.95	0.8	0.36	16	32.0	0	0	4,280	15,800	3,830	14,100	89.3	4,770
2	Subak Penebel	34.8	7.14	1.1	0.22	31	62.0	0	0	5,760	21,200	5,690	21,000	98.7	6,520
3	Subak Caguh	42.3	12.56	1.3	0.38	31	62.0	0	0	7,270	26,800	7,270	26,800	100.0	8,230
4	Subak Meliling Timpag	5.6	0.00	0.2	0.00	50	100.0	2	100.0	945	3,480	945	3,480	100.0	1,070
5	Subak Meliling	16.3	1.52	0.5	0.05	38	76.0	0	0	2,790	10,300	2,790	10,300	100.0	3,160
6	Subak Gadungan	18.3	0.87	0.6	0.03	44	88.0	1	50.0	3,230	11,900	3,230	11,900	100.0	3,650
7	Subak Sungsang	16.7	1.35	0.5	0.04	43	86.0	0	0	2,860	10,500	2,860	10,500	100.0	3,240
	Total	160.4	35.4	4.9	1.08					27,100	100,000	26,600	98,100	98.0	30,600
21 % of recoverable flow															
1	Subak Aya	26.5	11.95	0.8	0.36	16	32.0	0	0	4,280	15,800	3,830	14,100	89.3	4,770
2	Subak Penebel	34.9	10.77	1.1	0.33	20	40.0	0	0	5,760	21,200	5,600	20,600	97.1	6,520
3	Subak Caguh	42.2	21.43	1.3	0.65	19	38.0	0	0	7,270	26,800	7,270	26,800	100.0	8,230
4	Subak Meliling Timpag	5.6	0.00	0.2	0.00	50	100.0	2	100.0	945	3,480	945	3,480	100.0	1,070
5	Subak Meliling	16.6	4.82	0.5	0.15	31	62.0	0	0	2,790	10,300	2,800	10,300	100.0	3,160
6	Subak Gadungan	18.5	3.77	0.6	0.12	34	68.0	0	0	3,230	11,900	3,230	11,900	100.0	3,650
7	Subak Sungsang	17.2	5.06	0.5	0.15	37	74.0	0	0	2,860	10,500	2,860	10,500	100.0	3,240
	Total	161.5	57.8	4.9	1.76					27,100	100,000	26,500	97,700	97.7	30,600

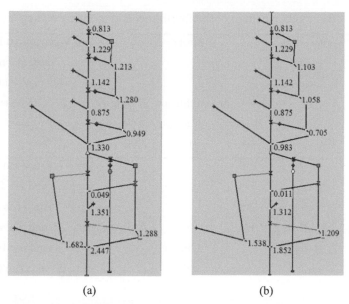

(a) (b)

Figure 7.29. Confluence flows for the third scenario at 62 and 21%

7.4.4   Simulation of the fourth scenario

In the fourth scenario the land preparation starts in period 16 (Aug II) for Subak Aya and in sequence for the downstream schemes, with 15 days following from the third scenario. The cropping pattern is shown in Table 7.10. Subak schemes upstream of TelagaTunjung Reservoir remain to require attention for the hydro water balance. Subak Aya (Figure 7.30) shows a different trend compared to the previous scenarios with deficit irrigation during Feb I - Apr I, May II, and during Sep I - Dec I. Subak Penebel also has a shifting trend from the previous scenarios with deficit irrigation at Jan I, during Feb II - May II, and during Sep II - Dec I (Figure 7.31).

Fluctuating trends occurred in Subak Caguh (Figure 7.32) with early in the year, Jan I deficit irrigation, as well as from Mar I until June II, and during Sep II - Dec I. Finally, Subak Meliling-Timpag still has no deficit irrigation (Figure 7.33).

The marks of confluence flows (Figure 7.34) which are typical in Subak Aya to Subak Penebel are 1.21 and 1.10 m^3/s, Subak Penebel to Subak Caguh 1.28 and 1.06 m^3/s, Subak Caguh to Subak Meliling-Timpag 0.96 and 0.71 m^3/s.

Table 7.10. The cropping pattern of the fourth scenario in Subak Agung Yeh Ho

Subak irrigation schemes	Functional paddy fields	Blocks			When to start land preparation Paddy I and Paddy II (24 periods per year)
		Upstream (*Ngulu*)	Midstream (*Maongin*)	Downstream (*Ngasep*)	
	ha	ha	Ha	ha	
1. Aya	644	644			Aug II (16) and Jan II (2)
2. Penebel	731	731			Sep I (17) and Feb I (3)
3. Riang	25	25			Sep II (18) and Feb II (4)
4. Jegu	111	111			Oct I (19) and Mar I (5)
5. Caguh	1093		1093		Feb II (4) and Sep II (18)
6. Meliling-Timpag	142		142		Mar I (5) and Oct I (19)
7. Telaga Tunjung Reservoir:					
• Meliling	420		420		Mar II (6) and Oct II (20)
• Sungsang	430			430	Mar II (6) and Nov II (22)
• Gadungan	485		485		Apr I (7) and Nov I (21)

In addition, downstream of Telaga Tunjung Reservoir, Subak Meliling has two confluence flows: to Yeh Ho River 0.07 and 0.02 m^3/s, and to Subak Sungsang 1.40 and 1.27 m^3/s. The results of the simulation of water allocation and agriculture production are shown in Table 7.11, with an average actual field level production of 98.9% for 62 and 21%.

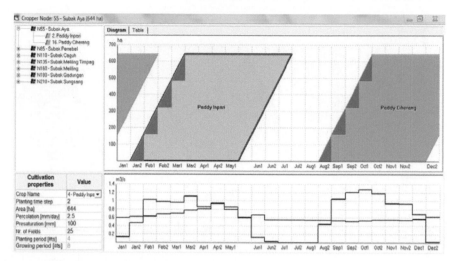

Figure 7.30. Cropping pattern and water balance of the fourth scenario for Subak Aya

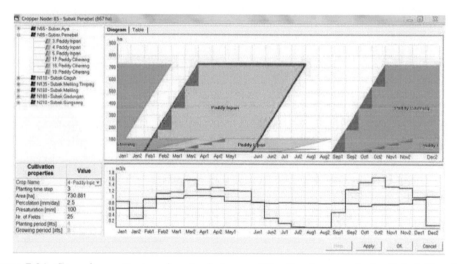

Figure 7.31. Cropping pattern and water balance of the fourth scenario for Subak Penebel

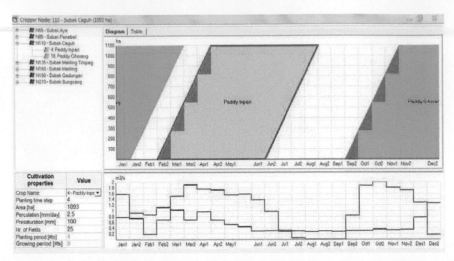

Figure 7.32. Cropping pattern and water balance of the fourth scenario for Subak Caguh

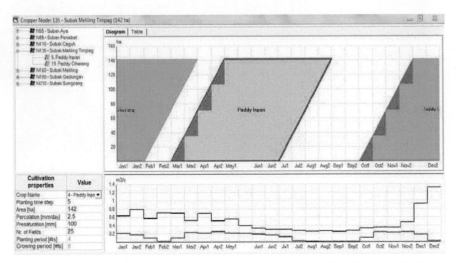

Figure 7.33. Cropping pattern and water balance of the fourth scenario for
Subak Meliling-Timpag

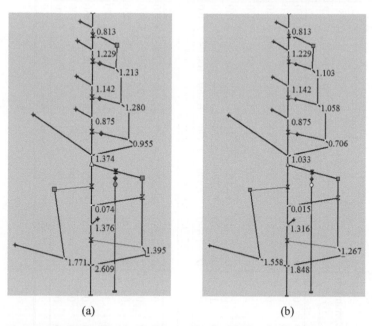

(a)   (b)

Figure 7.34. Confluence flows for the fourth scenario at 62 and 21%

7.4.5   Simulation of the fifth scenario

In the fifth scenario the land preparation starts in period 17 (Sep I) for Subak Aya and progresses for the downstream schemes, with 15 days following from the fourth scenario. The cropping pattern is shown in Table 7.12. Subak Aya (Figure 7.35) shows a fluctuating trend with deficit irrigation in Jan I, May I - II, and during Sep II - Dec I. Afterwards, Subak Penebel also shows a shifting trend compared to the previous scenarios with deficit irrigation in Jan I - II, during Mar I - Jun I, and Oct I - Dec I (Figure 7.36).

Table 7.11. Water allocation and agriculture production of the fourth scenario

Node	Name Subak Scheme	Yearly average				Success time steps		Success years		Pot. field level prod.		Act. field level prod.			Act. prod. costs
		Demand (MCM)	Shortage (m³/s)	Demand (MCM)	Shortage (m³/s)	number (-)	rate (%)	number (-)	rate (%)	Ton	10⁶ Rp	Ton	10⁶ Rp	(%)	10⁶ Rp
62 % of recovarable flow															
1	Subak Aya	26.1	11.71	0.8	0.36	17.0	34.0	0	0	4,280	15,800	3,980	14,600	92.8	4,850
2	Subak Penebel	34.4	6.86	1.0	0.21	30.0	60.0	0	0	5,760	21,200	5,760	21,200	100.0	6,520
3	Subak Caguh	42.0	11.87	1.3	0.36	29.0	58.0	0	0	7,270	26,800	7,270	26,800	100.0	8,230
4	Subak Meliling Timpag	5.5	0.00	0.2	0.00	50.0	100.0	2	100.0	945	3,480	945	3,480	100.0	1,070
5	Subak Meliling	15.7	1.24	0.5	0.04	41.0	82.0	0	0	2,790	10,300	2,790	10,300	100.0	3,160
6	Subak Gadungan	18.1	0.42	0.6	0.01	45.0	90.0	1	50.0	4,270	15,700	4,270	15,700	100.0	4,700
7	Subak Sungsang	16.7	1.61	0.5	0.05	45.0	90.0	0	0	1,890	7,000	1,890	6,960	100.0	2,080
	Total	158.5	33.7	4.8	1.03					27,200	100,000	26,900	99,000	98.9	30,600
21 % of recovarable flow															
1	Subak Aya	26.1	11.71	0.8	0.36	17.0	34.0	0	0	4,280	15,800	3,980	14,600	92.8	4,850
2	Subak Penebel	34.4	9.94	1.0	0.30	23.0	46.0	0	0	5,760	21,200	5,760	21,200	100.0	6,520
3	Subak Caguh	41.7	20.60	1.3	0.63	21.0	42.0	0	0	7,270	26,800	7,270	26,800	100.0	8,230
4	Subak Meliling Timpag	5.5	0.00	0.2	0.00	50.0	100.0	2	100.0	945	3,500	945	3,480	100.0	1,070
5	Subak Meliling	16.0	3.91	0.5	0.12	28.0	56.0	0	0	2,790	10,200	2,790	10,300	100.0	3,160
6	Subak Gadungan	18.2	3.31	0.6	0.10	35.0	70.0	0	0	4,270	15,700	4,270	15,700	100.0	4,700
7	Subak Sungsang	17.0	4.59	0.5	0.14	38.0	76.0	0	0	1,890	7,000	1,890	6,960	100.0	2,080
	Total	158.8	54.1	4.8	1.65					27,200	100,000	26,900	99,000	98.9	30,600

Table 7.12. The cropping pattern of the fifth scenario in Subak Agung Yeh Ho

Subak irrigation schemes	Functional paddy fields	Blocks			When to start land preparation Paddy I and Paddy II (24 periods per year)
		Upstream (*Ngulu*)	Midstream (*Maongin*)	Downstream (*Ngasep*)	
	ha	ha	ha	ha	
1. Aya	644	644			Sep I (17) and Feb I (3)
2. Penebel	731	731			Sep II (18) and Feb II (4)
3. Riang	25	25			Oct I (19) and Mar I (5)
4. Jegu	111	111			Oct II (20) and Mar II (6)
5. Caguh	1093		1093		Mar I (5) and Oct I (19)
6. Meliling-Timpag	142		142		Mar II (6) and Oct II (20)
7. Telaga Tunjung Reservoir:					
• Meliling	420		420		Apr I (7) and Nov I (21)
• Sungsang	430			430	Apr I (7) and Dec I (23)
• Gadungan	485		485		Apr II (8) and Nov II (22)

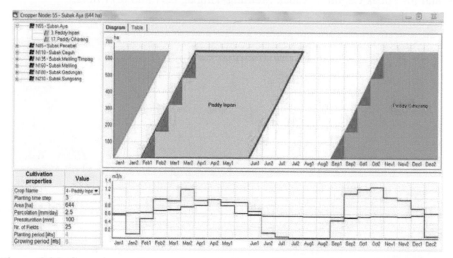

Figure 7.35. Cropping pattern and water balance of the fifth scenario for Subak Aya

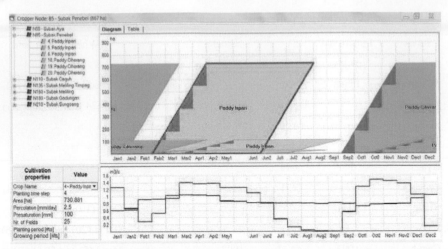

Figure 7.36. Cropping pattern and water balance of the fifth scenario for Subak Penebel

The fluctuating and shifting trends compared to the fourth scenario take place on Subak Caguh (Figure 7.37) with early in the year, Jan I - II, deficit irrigation, next from Mar II until July I, and during Oct I - Dec I. After all, Subak Meliling-Timpag still has no deficit irrigation (Figure 7.38).

The results of confluence flows for upstream schemes (Figure 7.39) for Subak Aya to Subak Penebel are 1.21 and 1.10 m³/s, Subak Penebel to Subak Caguh 1.29 and 1.06 m³/s, and for Subak Caguh to Subak Meliling-Timpag 0.97 and 0.71 m³/s.

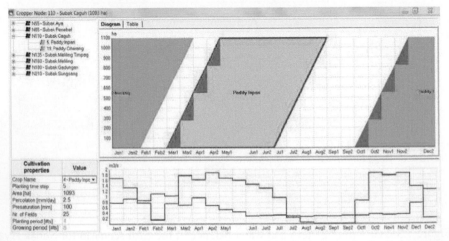

Figure 7.37. Cropping pattern and water balance of the fifth scenario for Subak Caguh

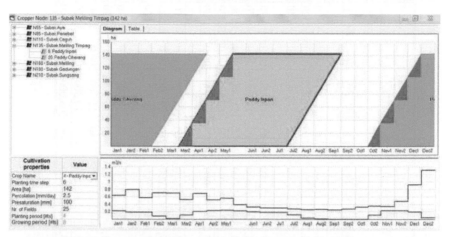

Figure 7.38. Cropping pattern and water balance of the fifth scenario for

Subak Meliling-Timpag

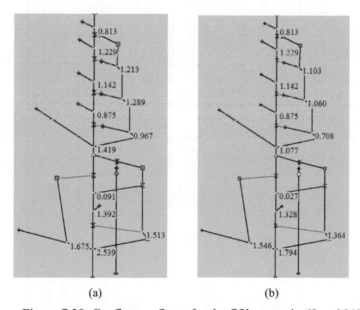

(a)   (b)

Figure 7.39. Confluence flows for the fifth scenario 62 and 21%

Furthermore, Subak Meliling has two confluence flows: to Yeh Ho River 0.09 and
0.03 m³/s, and to Subak Sungsang 1.51 and 1.36 m³/s. The results of the simulation of
water allocation and agriculture production are shown in Table 7.13, with an average
actual field level production of 100% for 62 and 21%.

Table 7.13. Water allocation and agriculture production of the fifth scenario

Node	Name Subak Scheme	Yearly average				Success time steps		Success years		Pot. field level prod.		Act. field level prod.			Act. prod. costs
		Demand	Shortage	Demand	Shortage	number	rate	number	rate						
		(MCM)	(m³/s)	(MCM)	(m³/s)	(-)	(%)	(-)	(%)	Ton	10⁶ Rp	Ton	10⁶ Rp	(%)	10⁶ Rp
62 % of recoverable flow															
1	Subak Aya	25.6	11.31	0.8	0.34	18	36.0	0	0	4,280	15,800	4,280	15,800	100.0	4,850
2	Subak Penebel	34.3	6.07	1.0	0.19	29	58.0	0	0	5,760	21,200	5,760	21,200	100.0	6,520
3	Subak Caguh	42.1	11.60	1.3	0.35	27	54.0	0	0	7,270	26,800	7,270	26,800	100.0	8,230
4	Subak Meliling Timpag	5.3	0.00	0.2	0.00	50	100.0	2	100.0	945	3,480	945	3,480	100.0	1,070
5	Subak Meliling	15.7	1.13	0.5	0.04	39	78.0	0	0	3,700	13,600	3,700	13,600	100.0	4,070
6	Subak Gadungan	18.0	0.51	0.5	0.02	47	94.0	0	0	3,180	11,700	3,180	11,700	100.0	3,400
7	Subak Sungsang	17.0	1.41	0.5	0.04	44	88.0	0	0	2,820	10,300	2,820	10,300	100.0	3,020
	Total	158.0	32.0	4.8	0.98					28,000	103,000	28,000	103,000	100	31,200
21 % of recoverable flow															
1	Subak Aya	25.6	11.31	0.8	0.34	18	36.0	0	0	4,280	15,800	4,280	15,800	100.0	4,850
2	Subak Penebel	34.5	9.79	1.1	0.30	23	46.0	0	0	5,760	21,200	5,760	21,200	100.0	6,520
3	Subak Caguh	42.4	20.67	1.3	0.63	22	44.0	0	0	7,270	26,800	7,270	26,800	100.0	8,230
4	Subak Meliling Timpag	5.3	0.00	0.2	0.00	50	100.0	2	100.0	945	3,480	945	3,480	100.0	1,070
5	Subak Meliling	15.8	3.96	0.5	0.12	28	56.0	0	0	3,700	13,600	3,700	13,600	100.0	4,070
6	Subak Gadungan	18.1	3.09	0.6	0.09	34	68.0	0	0	3,180	11,700	3,180	11,700	100.0	3,400
7	Subak Sungsang	17.3	4.75	0.5	0.15	38	76.0	0	0	2,820	10,300	2,820	10,300	100.0	3,020
	Total	159.1	53.6	4.8	1.63					28,000	103,000	28,000	103,000	100	31,200

7.5 Summary of the simulation and optimisation of Yeh Ho River Basin

The river basin simulation used the 80% dependable discharge, which is a standard for irrigation in Indonesia, as well as in the simulation and optimisation of the Subak schemes in Yeh Ho River Basin that was based on shifting of the start of land preparation (*nyorog*). This is one of the three activitees of Subak Associations related to operation and maintenance of Subak irrigation schemes. The results show that the fifth scenario resulted in an overall optimum agriculture production of 100% and a feasibility of farming (B/C) of 2.3 of actual field level production for the Subak irrigation schemes.

The fifth scenario starts in the period where the availability of the discharge of the upstream weirs, Aya and Penebel is stable (Figures 4.3 and 4.4). Compared to the other scenarios, especially with respect to the upstream schemes, September - October and February - March of the fifth scenario are the best periods to start land preparation. Because of the sequence that is followed by the Subak irrigation schemes, based on upstream (*ngulu*), midstream (*maongin*), and downstream (*ngasep*), the starting month of land preparation is also important for the allocation of water under deficit conditions.

Application of the right scheme model in RIBASIM for representing the Yeh Ho River Basin can offer optimum results of agriculture production, though there was limited and reduced water for irrigation from upstream sources and the river. The recoverable flow in the river basin scheme model played an important role for processing the simulation and optimisation of Yeh Ho River Basin.

7.5.1   Utilisation of hydraulic structures

Utilisation over the whole simulation period per diversion and of the reservoir are defined by the ratio between the sum of minimum (actual diverted and target diverted flow) over all downstream links and sum of upstream flow as shown in Table 7.14. The operation of the Telaga Tunjung Reservoir is quite important. The values of utilisation between 60.9 until 79.0% for all scenarios, and the graphs of reservoir operation are shown in Annex K. For the upstream weirs the values of utilisation between 91.3 until 93.4% are applicable.

Table 7.14. Utilisation of the hydraulic structures (%)

	First		Second		Third		Fourth		Fifth		Existing	
	62%	21%	62%	21%	62%	21%	62%	21%	62%	21%	62%	21%
Telaga Tunjung Reservoir	60.9	71.4	64.5	74.2	65.1	77.0	67.8	79.0	66.3	78.5	66.8	77.0
Weir Aya	92.2	92.2	92.2	92.2	92.2	92.2	92.2	92.2	92.2	92.2	92.2	92.2
Weir Penebel	91.3	92.2	91.7	93.1	92.5	93.4	92.4	93.4	92.0	92.8	91.3	92.2
Weir Caguh	86.1	87.1	85.7	88.3	86.6	88.6	86.1	88.4	85.5	86.8	86.1	87.1
Weir Meliling	74.4	75.4	74.2	76.5	75.0	76.8	74.6	76.7	74.1	75.2	74.4	75.4
Jetflow	11.7	18.6	11.8	18.5	11.7	18.9	11.7	19.5	12.0	19.7	11.5	17.2
Weir Gadungan	38.7	50.0	40.0	55.0	41.1	55.1	42.1	55.5	41.7	57.2	43.1	57.3
Weir Sungsang	15.3	30.1	19.1	33.9	19.0	34.1	18.9	35.5	20.0	37.3	15.7	32.1

7.5.2   Verification of the model

Besides the irrigation supply from the main river system, the recoverable flow from scheme to scheme, that has been used by the Subak farmers since a long time, is also essential. The results at the last node of recoverable flow in the model can be verified with the dependable flow of 80% over the last weir, namely Sungsang. Based on simulation of the five scenarios the verified results of the downstream flows show that the fifth scenario with 21% of recoverable flow gives the highest value of 0.77, as shown in Table 7.15 and Figure 7.40. All graphs of verification are shown in Annex K.

Table 7.15. R^2 main Yeh Ho Sungsang and recoverable flow at Sungsang Weir

Verification of downstream flows	First		Second		Third		Fourth		Fifth		Existing	
	62%	21%	62%	21%	62%	21%	62%	21%	62%	21%	62%	21%
R^2	0.60	0.68	0.69	0.72	0.66	0.74	0.67	0.76	0.66	0.77	0.56	0.68

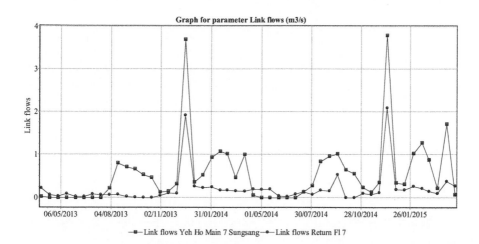

Figure 7.40. Downstream flows over Sungsang Weir for the fifth scenario with 21% of recoverable flow

8 Evaluation

8.1 Recommendations for river basin development

Especially in the upstream and midstream of Yeh Ho River the allocation of irrigation water to the Subak irrigation schemes experienced a deficit since the withdrawal of Gembrong Spring for domestic use and distribution of water from that spring was not based on applying the agreement on water sharing in practice. Therefor the first step in improved water supply would have to be to really apply the agreement about it in practice.

When the agreement will be applied in practice effective operation of the Telaga Tunjung Reservoir does not necessarily completely solve the problem of water shortage in the river basin. However, when a cropping pattern based on the investigated fifth scenario will be applied the water deficit and yield reduction will be quite limited. Based on simulation with the RIBASIM model the results of this study can provide a contribution to the management of water for the Subak irrigation schemes along Yeh Ho River.

In addition, the purpose of this research was to find answers on the gap between the theory and the practice that is being applied in the Subak irrigation schemes in Yeh Ho River Basin. Based on literature review, fieldwork, laboratory analyses, the standards book in Indonesia, and application of the RIBASIM model this research has brought the theory and practice closer, especially by the hydrological and hydraulic analysis.

The results of the study also show that the role of recoverable flow is very important in the water distribution from upper schemes to lower schemes. Subak farmers have the ability of negotiation based on the *THK* philosophy. The alternative scenarios by shifting the start of land preparation (*nyorog*) are expected to be well received by these farmers.

In a discussion with the leader of Subak Agung Yeh Ho and Subak farmers based on the optimized results on agriculture production of the fifth scenario, with land preparation starting in September - October and February - March in the upstream schemes: Aya and Penebel and similar changes in the start of land preparation of the midstream and downstream schemes, they agreed with the results.

The importance of this research is that an appropriate of scenario has been identified to resolve the issue of sustainable water management within Yeh Ho River Basin and the indigenous Subak irrigation schemes. The study outcomes also show that, even though before this study it was never evidenced quantitatively at river basin scale, the Subak farmers are able to manage successfully their irrigation systems, based on the local wisdom. Furthermore, at river basin scale, the results of the study can provide a contribution for similar Subak irrigation schemes in other river basins in Indonesia.

8.1.1   *Telaga Tunjung Reservoir operation based on Subak cropping patterns*

The subsistence of the reservoir has been applied as important role to increase water supply for downstream schemes: Meliling, Gadungan and Sungsang. It will also have an direct impact to sustain water conservation at the river basin scale. Water conservation in this research is an effort of water resources management by using simulation on the availability of discharge in order to increase the efficiency of water demand. Based on the ratio between the actual diverted flow and the target diverted flow under the different scenarios the water demand in Yeh Ho River Basin was served by 60.9 - 79.0%. This was mainly caused by deficit in the supply of the systems upstream of the Telaga Tunjung Reservoir, while the operation of the reservoir resulted in fully the required supply of the downstream systems. This result can deliver good information to the Regional River Office of Bali-Penida. It is envisaged that they will consider this result under their authority related to the standard operation of Telaga Tunjung Reservoir.

The reservoir can be operated to allocate the irrigation water in an optimum way by connecting to the river basin and the scenarios of the Subak schemes cropping patterns. From the cropping pattern of the first scenario - which reflects the present situation - shifting of the start of land preparation each time with 15 days is the most important action to optimise agriculture production in Yeh Ho River Basin from upstream, midstream and downstream schemes. This implies that for the fifth scenario land preparation starts 60 days later than in the first scenario.

Since the start of the operation in 2006, the operation staff of Telaga Tunjung Reservoir and Subak farmers have been dealt gradually that downstream systems are irrigated alternately from Telaga Tunjung Reservoir, one year to Meliling scheme, the next one year to Gadungan scheme. The last, Sungsang scheme is irrigated generally from recoverable flows. However the problem water sharing still occurs in the upstream and midstream schemes.

The component of *THK* on harmony among people could be applied to shift the overall start of land preparation in the schemes in accordance with the cropping pattern of the fifth scenario, so that the agricultural productivity in the schemes can be guaranteed in an optimal way. As well, it could be applied in relation to the reservoir operation to supply Subak irrigation schemes that can best be operated, especially for the downstream schemes: Meliling, Gadungan, and Sungsang.

8.1.2   Operation and maintenance of the Subak irrigation systems

The operation and maintenance technologies are linked to sub-section 3.2.1 about PIM in irrigation system operation and maintenance, and the third element of the *Tri Hita Karana* philosophy as shown in Table 3.1 on the material subsystem. One part the material subsystem concerning the natural relation elements, mentions that the presence of hydraulic structures and irrigation systems are suitable with the needs of the local farmers. Therefore the operation of the diversion weirs has been implemented based on a compromise between the irrigation observer and the Subak farmers related to the opening and closing of the gates of the weirs and the maintenance of the irrigation canals. For irrigation water supply the gates are opened by setting them at a point gauge based on the irrigated area (*daerah irigasi*). During maintenance the irrigation observer closes the gates of the secondary canals. However, while there are no gates in the distribution boxes within the Subak irrigation schemes, water can still flow through these boxes related to the natural hilly topography, the specific location of water sources and type of soils. Subak farmers make together a deal in *Awig-awig* Subak about opening and closing the irrigation water supply that only applies to the WDUs (*tektek*). Additionally this routine has been expressed in Table 3.1 related to the social subsystem and the material subsystem especially on the religious elements. Also the system of one inlet and one outlet enforces Subak irrigation schemes to apply the principle of justice in the management that has been adjusted to the natural environment and has proven to be an effective system.

Subak irrigation farmers would accept the new operation and maintenance strategy based on the revised start of land preparation of the fifth scenario, and through the results of the simulation on confluence flows the scheme model can provide information to Subak associations and Subak farmers per scheme about the supply of irrigation water that they can receive at the time of the start of land preparation. However, by approach with the RIBASIM model, the improvement of the agricultural production has been verified quantitatively in all the schemes in Yeh Ho River Basin.

9   References

Abernethy, B.R. (2002). The New Weibull Handbook. 536 Oyster Road, North Palm Beach, FL. 33408-4328 * 561-482-4082.

Acreman, M. (2000). Managed flow releases from reservoirs: issues and guidance, available at: http://www.dams.org/docs/kbase/contrib/ opt056.pdf

Akintug, B. (2010). Water resources engineering reservoirs: Ripple diagram method (1883), METU Northern Cyprus Campus, Cyprus.

Alemu, E.T., Palmer, R.N., Polebitski, A. and Meaker, B. (2011). Decision Support System for Optimizing Reservoir Operations Using Ensemble Streamflow Predictions. Journal of Water Resources Planning and Management 137:72–82.

Alexander, K.S., M. Moglia and C. Miller. (2010). Water needs assess: Learning to deal with scale, subjectivity and high stakes, Journal of Hydrology 388 (2101) 251-257.

Ali, M.D.H. and Shui, L.T. (2001). Optimal Allocation of Monthly Water Withdrawals in Reservoir Systems. http://dx.doi.org/10.1023/ A:1014467516914

Allen, R.G., Pereira, L.S., Raes, D. and Smith, M. (1998). Crop evapotranspiration - Guidelines for computing crop water requirements - FAO irrigation and drainage paper 56. FAO (Food and Agriculture Organization of the United Nations), Rome, Italy.

Anjasari, L. (2015). Constitutional Court. http://www.mahkamahkonstitusi.go.id/ index.php?page=web. Berita&id =10634#.VfeZsGfovIU

Arga, I.W. (2011). Awig-awig Subak as a tool of management Subak Association: case Subak Lanyahan at Krobokan, Buleleng, Bali. The Excellence Research of Udayana University, Bali.

Arief, S.S. (1999). Applied philosophy of Tri Hita Karana in design and management of Subak irrigation system, UGM, Yogyakarta, Indonesia.

Arsana, I G.K.D. (2012). The Efficiency of paddy consumptive use in the upstream, midstream and downstream of Yeh Ho River Basin, Bali. Doctoral dissertation: Graduate Program of Gadjah Mada University, the Faculty of Agriculture, Yogyakarta, Indonesia.

Asnawi, R. (2015). Climate change and food sovereignty in Indonesia. Sosio Informa Vol. 1, No:293–309.

Asfaw, T. D., and Saiedi, S. (2011). Optimal short-term cascade reservoirs operation using genetic algorithm. Asian Journal of Applied Sciences.

Baskoro, D. P. T., and Tarigan, S. D. 2007. Soil Moisture Characteristics on Several Soil Types. Jurnal Tanah dan Lingkungan 9:77–81.

Barker, R. and Molle, F. (2004). Irrigation management in rice-based cropping systems: issues and challenges in Southest Asia. http://www.agnet.org/library/eb/543/

Berris, S.N., Hess, G.W. and Bohman, L.R. (1998). River and reservoir operation model, Truckee River Basin, California and Nevada. Water-Resources Investigations Report 01-4017.

Bimantara, J. G. (2015). The old law safe the situation. http://print.kompas.com/ baca/sains/lingkungan/2015/03/12/Undang-undang-Lama-Selamatkan-Situasi

Birkelbach, A.W., Jr. (1973). The Subak Association. http://cip.cornell.edu/ seap.indo/1107129338

Bokings, D. L., Sunarta, I N., and Narka, I W. (2013). *Karakteristik Terasering Lahan Sawah dan Pengelolaannya di Subak Jatiluwih, Kecamatan Penebel.* 2:175–183.

Borden, C., Gaur, A., and Singh, C. R. (2016). Water Resource Software Application Overview and Review (The World.). SAW South Asia Water Initiative.

Bridgewater, P. and Bridgewater, C. (2003). Is there a future for cultural landscapes?. Available at: http://edepot.wur.nl/119326

Bruin, H.A.R.D.E. 1983. Evapotranspiration in humid tropical regions.

Case, M., F. Ardiansyah and E. Spector. (2008). Climate change in Indonesia implications for humans and nature http://www.worldwildlife.org/ climate/Publications/ WWFBinaryitem7664.pdf

Chen, F.W., Liu, C.W. and Chang, F.J. (2014). Improvement of the agricultural effective rainfall for irrigating rice using the optimal clustering model of rainfall station network. Paddy and Water Environment 12:393–406.

Chow, Ven T., Maidment, D.R. and Mays, L.W. (1988). Applied Hydrology. International Editions. McGraw-Hill Book Co. – Singapore. ISBN 0-07-100174-3.

Coward, Jr. and Levine, G. (1987). Studies of farmer-managed irrigation systems: Ten years of cumulative knowledge and changing research priorities. Ed IIMI/WECS/Government of Nepal, 1-31. Colombo, Sri Lanka.

D'Ambra, A., Crisci, A., and Sarnacchiaro, P. (2015). A generalized analysis of the dependence structure by means of ANOVA. Journal of Applied Statistics.

Department of Internal Affairs. (2004). Recapitulation the numbers of Indonesia Islands in 2004. Available at http://wapedia.mobi/id/ Jumlah_pulau_di_Indonesia

Department of Public Works. (1986). KP - 01 Criteria for Planning-The Planning Irrigation Network.

Department of Public Works. (1986). PT - 01 Technical Requirements - The Planning Irrigation Network.

Department of Public Works. (2004). Project of Bali Irrigation. Final Report.

Department of Public Works. (2005). Irrigation Schemes in Yeh Ho River Basin. Irrigation Bali Project.

DHV Consutants BV and Delft Hydraulics. (1999). Hydrology Project Training Module: How to Analyse Discharge Data. Version Nov. 99.

Emerson, W., and Foster, R. (1985). Aggregate classification and soil physical properties for rice-based cropping systems (CSIRO Div.). Adelaide, South Australia.

Falvo, D. (2000). On Modeling Balinese Water Temple Networks as Complex Adaptive. Systems. Human Ecology , Volume 28, Issue 4, pp 641-649.

Food and Agriculture Organization of the United Nations (FAO). (2005). AQUASTAT water availability information by country. Available at http:// www.greenfacts.org/ en/water-resources/figtableboxes/3.htm

Food and Agriculture Organization of the United Nations (FAO). (2006). World water resources by country. Available at http://www.fao.org/DOCREP/005/Y4473E/ y4473e0g.gif

Fowler, P.J., (2003). World Heritage Cultural Landscapes, 1992–2002: a Review and Prospect. World Heritage Paper 7: Cultural Landscapes: the Challenges of Conservation, World Heritage 2002, Shared Legacy, Common Responsibility, Associated Workshops, 11-12 November 2002, Ferrara - Italy. Published in 2003 by UNESCO World Heritage Centre.

Fowler, P.J. (2003). World heritage paper 6 World heritage cultural landscapes 1992-2002. Published by UNESCO World Heritage Centre.

Fukamachi, K., Oku, H. and Miyake, A. (2005). The relationships between the structure of paddy levees and the plant species diversity in cultural landscapes on the west side of Lake Biwa, Shiga, Japan. Landscape Ecol Eng (2005) 1: 191–199 DOI 10.1007/s11355-005-0019-8.

Gany, A.H.A. (2004). Subak irrigation system in Bali in irrigation history of Indonesia. Ministry of Settlement and Regional Infrastructures in Collaboration with: The Indonesian National Committee of International Commission on Irrigation and Drainage-ICID. ISBN: 979-96442-3-2.

Graders, C.J. (1939). Subak in Jembrana Kingdom. Biro Dokumentasi dan Publikasi Hukum dan Pengetahuan Masyarakat, Udayana University, 1979.

Groenfeldt, D. (2005). Irrigation development and indigenous people. Available at: http://www.iwmi.cgiar.org/Assessment/index.htm

Hanks, R. J., and Ashcroft, G. L. (1980). Applied soil physics: advanced studies agricultural science. Springer, Berlin.

Hardjowigeno, S. (2003). Soil classification and pedogenesis. Medyatama Sarana Perkasa, Jakarta.

Hauser-Schäublin, B. (2011). Land Donations and the Gift of Water. On Temple Landlordism and Irrigation Agriculture in Pre-Colonial Bali. Human ecology: an interdisciplinary journal 39:43–53.

Hejazi, M.I. and Cai, X. (2011). Building more realistic reservoir optimization models using data mining – A case study of Shelbyville Reservoir. Advances in Water Resources 34:701–717.

Hillel, D. (1998). Environmental soil physics: Fundamentals, applications, and environmental considerations. Academic Press, USA.

Huang, Y., Li, Y.P., Chen, X. and Ma, Y.G. (2012). Optimization of the irrigation water resources for agricultural sustainability in Tarim River Basin, China. Agricultural Water Management 107:74–85.

International Union for Conservation of Nature (IUCN). (2003). Flow; The essentials of environmental flows. Available at: http://www.iucn.org/ dbtw-wpd/edoc.s/2003-021.pdf

Janssen, A.M. (2007). Coordination in irrigation systems: An analysis of the Lansing–Kremer model of Bali. Agricultural Systems, Volume 93, Issues 1ww, March 2007, Pages 170scien

Jayadi, R. and Darmanto. (2011). Presentation: Challenges of integrated water resources management in developing country, case Indonesia, at United Nation University, Japan.

Jha, N. and Schoenfelder, J. (2011). 'Studies of the Subak: new directions, new challenges.' Human Ecology 39(1): 3-10.

Johansson, R.C. (2000). Pricing Irrigation Water: A Literature Survey. Policy Research Working Paper. The World Bank Rural Development Department. Washington DC, USA.

Johansson, R.C., Tsur, Y., Roe, T.L. and Doukkali, R. (2002). Pricing irrigation water: a review of theory and practise. Water Policy 4 (2002) 173-199. Elsevier Science Ltd.

Kaler, S.P. (1985). Subak: Socio-cultural overview. HATHI Seminar, Denpasar, Indonesia.

Kang, M. and Park, S. (2014). Modeling water flows in a serial irrigation reservoir system considering irrigation return flows and reservoir operations. Agricultural Water Management 143:131–141.

Kim, T.C., Lee, J.M. and Kim, D.S. (2003). Decision support system for reservoir operation considering rotational supply over irrigation blocks. Paddy Water Environ 1:139-147.

Kim, H.K., Jang, T.I., Im, S.J. and Park, S.W. (2009). Estimation of irrigation return flow from paddy fields considering the soil moisture. Agricultural Water Management 96 (2009) 875-882.

Kiyoumarsi, F. (2015). Mathematics Programming based on Genetic Algorithms Education. Procedia - Social and Behavioral Sciences 192:70–76.

Lansing, J.S. (1991). Priest and Programmers: Technologies of Power in the Engineered Landscape of Bali. New Jersey: Princeton University Press.

Lansing, J.S. (1987). Balinese water temples and the management of irrigation. American Anthropologist 89.

Larson, W. E., and Clapp, C. E. (1984). Effects of organic matter on soil physical properties, in Organic Matter and Rice. Phillipines.

Loebis, J. (2003). Towards a new paradigm for integrated water resources management and development in Indonesia. Water Resources Systems-Hydrological Risk, Management and Development. IAHS Publ. no 281, 2003.

Lorenzen, R.P. (2015). Disintegration, Formalisation or Reinvention? Contemplating the Future of Balinese Irrigated Rice Societies. The Asia Pacific Journal of Anthropology 16:176–193.

Lorenzen, R.P. and Lorenzen, S. (2011). Changing Realities - Perspectives on Bali Rice Cultivation. Hum Ecol (2011) 39:29-42. Doi 10.1007/s10745-010-9345-z

Lorenzen, R.P. and Lorenzen, S. (2005). A case study of Balinese irrigation management: institutional dynamics and challenges. Full paper for the 2nd Southeast Asian Water Forum, 29 August - 3 September 2005, Bali, Indonesia.

Luchman, H., Kim, J.E., and Hong, S.K. (2009). Cultural landscape and ecotourism in Bali Island, Indonesia. Journal Ecol. Field Biol. 32 (1): 1-8.

MacRae, G. and Arthawiguna, I. (2011). 'Sustainable agricultural development in Bali: Is the Subak an obstacle, an agent or subject?' Human Ecology 39(1): 11-20.

McMahon, T.A., Pegram, G.G.S., Vogel, R.M. and Peel, M.C. (2007). Revisiting reservoir storage–yield relationships using a global streamflow database. Advances in Water Resources, Volume 30, Issue 8, August 2007, Pages 1858.

Md. Azamathulla, H., Wu, F.C., Ghani, A.A., Narulkar S.M., Zakaria, N.A. and Chang, C.K. (2008). Comparison between genetic algorithm and linear programming approach for real time operation. Journal of Hydro-environment Research 2:172–181.

Mousavi, S.J., Ponnambalam, K. and Karray F. (2003). Deriving reservoir operation rules via Fuzzy Regression and ANFIS.

Mugatsia, E.A. (2010). Simulation and scenario analysis of water resources management in Perkerra catchment using Weap Model. Department Of Civil And Structural Engineering, School Of Engineering Of MOI University, Kenya.

Naylor, R. (2007). Assessing risks of climate variability and climate change for Indonesian rice agriculture. PNAS Early Edition May 1, 2007.

Nishimura, S., Lee, S., Ito, K., and Senge, M., (2005). The mode of operation of a regulating reservoir for effective use of river flow. Paddy Water Environ 3: 149-154.

Norken, I.N., Suputra, I K. and Kertaarsana, I G.N. (2010). The History and Development of Sedahan as a Coordinator of Water Management for Subak in Bali (L'Histoire et le Développement de Sedahan en tant que Coordonnateur de la Gestion de l'eau pour Subak à Bali). In: Proceedings International Commission on Irrigation and Drainage (ICID-CIID) Yogyakarta, Indonesia.

Norken, I N., Windia, W. and Mudina I M. (2010). Effort to promote Subak as an irrigation commission in Bali. ICID CIID Yogyakarta, Indonesia

Olivares, M.A. (2008). Optimal hydropower reservoir operation with environmental requirements. Dissertation PhD in University of California, USA.

Omar, M.M. (2013). Evaluation of actions for better water supply and demand management in Fayoum, Egypt using RIBASIM. Water Science 27:78–90.

Othman A.K.A. and Mekhaizim, H.A.A. (2010). Harmonics elimination in multilevel inverter using linear Fuzzy regression. World Academy of Science, Engineering and Technology 62.

Pedersen, L. and Dharmiasih, W. (2015). The Enchantment of Agriculture: State Decentering and Irrigated Rice Production in Bali. The Asia Pacific Journal of Anthropology 16:141–156.

Pedro-Monzons, M., Ferrer, J., Solera, A., Estrela, T., and Paredes-Arquiola, J. (2015). Key issues for determining the exploitable water resources in a Mediterranean river basin. Science of the Total Environment 503-504:319–328.

Pereira, L.S., Allen, R.G., Smith M., and Raes, D. (2015). Crop evapotranspiration estimation with FAO56: Past and future. Agricultural Water Management 147:4–20.

Petkovsek, G. and Roca, M. (2014). Impact of reservoir operation on sediment deposition. Proceedings of the ICE - Water Management 167:577–584.

Perry, C. (2007). Efficient Irrigation; Inefficient Communications; Flawed Recommendations. Irrig. and Drain. 56: 367-378 (2007). Published online in Wiley InterScience. DOI: 10.1002/ird.323

Poffenberger, M. and Zurbuchen, M.S. (1979). The economics of village Bali: Three perspectives. Post-doctoral research in University of Barkeley and University of Michigan, USA.

Pribadi and Wena, M. (2007). Aspect cultural studies of technology and canal model and building Subak irrigation: a case study in Subak organization in Tabanan - Bali. Educational Research Journal 17, No. 2/Dec 2007, State University of Malang, Indonesia.

Pusposutardjo, S. (1997). Insights (Vision) Linkages With the Future of Irrigation Water Resources Management. The papers presented at the Regional Workshop on Water Empowerment, Directorate General of Irrigation, Bali, Indonesia.

Rachman, H.P.S. and Purwoto, A. (2005). Management policies food reserve in the regional autonomy and PERUM BULOG. *Forum Penelitian Agro Ekonomi* 23:73–83.

Raka, A.A. (2009). Land-field in Bali Island.

Rashid, M.U., Shakir, A.S., and Khan, N.M. (2014). Evaluation of Sediment Management Options and Minimum Operation Levels for Tarbela Reservoir, Pakistan. Arabian Journal for Science and Engineering 39:2655–2668.

Regional River Office of Bali-Penida, (2006). Detail Design Telaga Tunjung Dam. Final Report.

Regional River Office of Bali-Penida, (2011). Profile of central river region Bali-Penida. Available at http://www.pu.go.id/satminkal/dit_sda/ profil%20balai/ bws/ profilebalaibalipenida.pdf

Romenah. (2010). Potential land and critical land. Module: Geography, Jakarta, Indonesia.

Roth, D., and Sedana, G. (2015). Reframing Tri Hita Karana : From 'Balinese Culture' to Politics. The Asia Pacific Journal of Anthropology 16:157–175.

Roth, D. (2011). The Subak in diaspora: Balinese farmers and the Subak in South Sulawesi. Hum Ecol (2011) 39:55–68. DOI 10.1007/s10745-010-9374-7

Saptaningsih, E., and Hariyanti, S. (2015). *Kandungan selulosa dan lignin berbagai sumber bahan organik setelah dekomposisi pada tanah Latosol. Buletin Anatomi dan Fisiologi XXIII.*

Statistical Central Agency. (2009). Average values of meteorological and geophysical condition by station. Available at http://bali.Central Bureau of Statistics.go.id/tabeldetail.php?ed=51000104&od= 1&rd=31&id=1

Statistical Central Agency. (2009). Meteorological and Geophysical Condition of Bali by Regencies. Available at http://bali.Central Bureau of Statistics.go.id/ tabeldetail.php?ed=51000105&od= 1&rd=31&id=1

Statistical Central Agency. (2009). Area and type of land use. http://bali.bps.go.id/ eng/tabel_detail.php?ed=607001&od=7&id=7

Statistical Central Agency, (2010). Harvest productivity-production of rice in Province of Bali. Available at http://www.Statistical Centre Agency.go.id/ tnmn_pgn.php?eng=0

Statistical Central Agency. (2011). Harvested Area, Production of Wetland and Dryland Paddy by Regency/City in Bali. http://bali.bps.go.id/ eng/tabel_detail.php? ed=607002&od=7&id=7

Statistical Central Agency Bali Province, (2013). http://www.aktual.co/ekonomibisnis/ 113106penjualan-gabah-kualitas-rendah-meningkat-di-bali

Straub, S. (2011). Water conflicts among different user groups in South Bali, Indonesia. Hum Ecol (2011) 39:69-79. DOI 10.1007/s10745-011-9381-3.

Sudjarwadi (1992). River basin development, page 80 PAU IT-UGM, Yogyakarta, Indonesia.

Suanda, D.K. and Suryadi, F.X. (2010). Bali's Subak water management systems in the past, present, and future. In: Proceedings International Commission on Irrigation and Drainage (ICID-CIID) Yogyakarta, Indonesia.

Sulis, A. and Sechi, G.M. (2013). Comparison of generic simulation models for water resource systems. Environmental Modelling & Software 40:214–225.

Sumiyati, Windia, W., Tika, I W., and Sulastri, N.N. (2013). Application the System Rice Intensification (SRI) uses intermittent irrigation techniques (ngenyatin) in development of rice productivity of Subak irrigation. National Seminar on KNI-ICID, 30 November 2013, Semarang, Indonesia.

Sunarta, I N. (2016). Latosol soil at Tabanan Regency. Bali, Indonesia.

Surata, S.P.K., (2003). Budaya padi dalam Subak sebagai model pendidikan lingkungan. Yayasan Padi Indonesia, Jakarta.

Sutawan, N. (2008). Organization and management of Subak in Bali. Pustaka Bali Post. ISBN 978-979-8496-73-8.

Sutawan, N. (2010). The existence of Subak Bali: Could be faced many challanges. Udayana University, Bali, Indonesia.

Trajkovic, S. and Gocic, M. (2010). Comparison of some empirical equations for estimating daily reference evapotranspiration. Facta universitatis - series: Architecture and Civil Engineering 8:163–168.

United Nations Educational, Scientific and Cultural Organization (UNESCO). (2006). Scientific facts on water resources. Available at http://www.greenfacts.org/en/water-resources/figtableboxes/8.htm

University of Texas Directory (UTDIRECT). (2002). Perry-Castañeda library map collection. The University of Texas at Austin. Available at http://www.lib.utexas.edu/maps/middle_east_and_asia/indonesia_rel_2002.pdf

Van der Krogt, W.N.M. (2013). RIBASIM Version 7.00. River Basin Simulation Model User Manual. Available at: www.wldelft.nl/soft/ribasim

Van der Krogt, W.N.M. (2008). RIBASIM Version 7.00 Technical Reference Manual.

Von Droste, B., Plachter, H. and Rossler, M. (1995). Cultural Landscapes of Universal Value. Components of a Global Strategy. Gustav Fisher Verlag, Stuttgart - New York, USA.

Walker, W.R. and Skogerboe, G.V. (1987). Surface Irrigation Theory and Practice. Prentice-Hall, Inc., Englewood Cliffs, New Jersey, USA.

Wapedia. (2011). Bali. Available at http://wapedia.mobi/id/Bali#1

Whitten, T., Soeriaatmadja, R.E. and Afif, S.A. (1996). The ecology of Java and Bali. The Ecology of Indonesia Series II. Periplus, Singapore.

Wikipedia. http://id.wikipedia.org/wiki/Tri_Hita_Karana

Windia, W. (2013). *Penguatan Budaya Subak Melalui Pemberdayaan Petani*. Journal of Bali Studies 03/02:137–158.

Windia W., Pusposutardjo, S., Sutawan, N., Sudira, P. and Arif, S.S. (2010). Subak irrigation system transformation based on the concept of Tri Hita Karana. In: Proceedings International Commission on Irrigation and Drainage (ICID-CIID) Yogyakarta, Indonesia.

Windia, W., Pusposutardjo, S., Sutawan, N., Sudira, P. and Arif, S.S. (2005). (THK) *Sebagai Teknologi Sepadan dalam Pertanian.* SOCA (Socio-Economic of Agriculturre and Agribusiness), 5(3).

World Meteorological Organization and Global Water Partnership/Integrated Flood Management. (2008). Reservoir operations and managed flows. Associated Programme on Flood Management. Water Resources Environment Technology Centre (WEC), Japan.

Wurbs, R.A. (2005). Modeling river/reservoir system management, water allocation, and supply reliability. Journal of Hydrology 300:100–113.

Wurbs, R.A. (1996). Modelling and analysis of reservoir system operations, page 260-288, Prentice Hall PTR, Upper Saddle River, New Jersey, USA.

Yekti, M.I., Schultz, B. and Hayde, L. (2012). Challenge of runoff regulation to supply paddy terraces in Subak irrigation schemes. In: Proceedings 7th Asian Regional Conference of International Commission on Irrigation and Drainage (ICID), 24-28 June 2012, Adelaide, Australia.

Yekti, M.I., Schultz, B. and Hayde, L. (2013). Development of a conceptual approach to manage flow of Subak irrigation schemes in Bali, Indonesia. In: Proceedings 1st World Irrigation Forum (WIF1), 29 September - 3 October 2013, Mardin, Turkey.

Yekti, M.I., Gany, A.H.A. and Schultz, B. (2014). Learning from decades of experience with Subak ancient Participatory Irrigation Management in Bali. Lecture note at Water and Land Management Institute (WALMI), 21 - 24 January 2014, Aurangabad, Maharashtra, India.

Yekti, M.I., Norken, I N., B. Schultz, B. and Hayde, L. (2014). A Role Concept of Reservoir Operation for Sustainable Water Supply to Subak Irrigation Schemes: Case Study of Yeh Ho River Basin. In: Proceeding International Symposium on DAMs in A Global Environmental Challenges. The 82th Annual Meeting of ICOLD (International Commission on Large Dam), 1 – 6 June, 2014, Bali, Indonesia.

Yekti, M.I., Schultz, B., Norken, I N., Gany, A.H.A. and Hayde, L. (2014). Irrigation-drainage (irrigation-drainage) of Subak Irrigation Schemes: A Farmer's Perspective over A Thousand Years. In: Proceeding 12 ICID International Drainage Workshop (IDW), 23 -26 June 2014, Saint-Petersburg, Russia.

Yulia. (2015). *18 Jenis Jenis Tanah di Indonesia : Manfaat, Persebaran, Gambarnya.*

Zahraie, B. and Hosseini, S.M. (2007). Development of fuzzy reservoir operation policies using genetic algorithm. IASME/WSEAS International Conference on Water Resources, Hydraulics and Hydrology, Slovenia.

Zahraie, B. and Hosseini, S.M. (2009). Development of reservoir operation policies considering variable agricultural water demands. Expert Systems with Applications 36 (2009) 4980-4987.

Zotarelli, B. L. and Dukes, M. D. (2010). Interpretation of Soil Moisture Content to Determine Soil Field Capacity and Avoid Over Irrigation in Sandy Soils Using Soil Moisture Measurements. Agricultural and Biological Engineering Department.

APPENDICES

Annex A. Abbreviations

BULOG	Bureau of logistics
BUMD	Regional owned corporation
FAO	Food and Agriculture Organization of the United Nations
FC	Field capacity
IFM	Integrated flood management
IUCN	International Union for Conservation of Nature and Natural Resources
IWRM	Integrated water resources management
MDG	Millennium development goals
MSL	Mean sea level
PDAM	Local water supply utility
PIM	Participatory irrigation system management
PMF	Probable maximum flood
PWP	Permanent wilting point
RIBASIM	River basin simulation model
Rp.	Rupiah (Indonesian currency)
TARWR	Total actual renewable water resources
THK	*Tri Hita Karana*
UNESCO	United Nations Educational, Scientific and Cultural Organization
VMF	*Voedings Middelen Fonds*
WDU	Water distribution unit
WMO	World Meteorological Organization
WTP	Water treatment plant

Annex B. Symbols

A Area under the curve with the proportion of the pre-saturated area versus time

A_i Average surface area of the reservoir in time interval i

A_t Values are specified in Δt

$A_{t+\Delta t}$ A function of volume of storage at the beginning and end of a time interval

B/C Feasibility of farming, which is determined by the difference of revenue and cost devided by the cost

B Benefit (Rp.)

C Cost (Rp.)

c Crop coefficient

C_p Crop factor

e Base of natural logarithms

er Evaporation rate (mm/day)

E_i Evaporation in the reservoir each time to i

E_{To} Reference evapotranspiration (mm/day)

E_{tp} Estimation of potential evapotranspiration (mm/day)

Gs Specific gravity

Kv Hydraulic conductivity (cm/hour)

m The rank of annual series data arranged in descending order of magnitude

M The sum of evapotranspiration and percolation (mm/day)

n Number of years

n Porosity

P Probability of exceedence (%)

Q_{inflow} Inflow (m^3/s)

Q_{divert} Diversion (m^3/s)

$Q_{overflow}$ Overflow (m^3/s)

$Q_{confluence}$ Confluence flow (m^3/ s)

R Revenue (Rp.)

R_e Effective rainfall (mm/day)

R^2 Percent of variance explained/coefficient of determination (%)

Req_{tot}	Total crop water requirement (mm/time step)
R80	80% probability of dependable rainfall (mm/day)
R/C	Efficiency of farming, which is determined by revenue divided by the cost
S	Water requirement for the pre-saturation of the field and water layer (mm)
S	Percolation (mm)
Sr	Saturation
t	Time (days)
t_0	Starting time of the time period between t_0 and t
T	Length of pre-saturation during the planting period (days)
X	Plotting position
X_m	Plotting position m
Y	area which has received the pre-saturation and water layer requirement and receives compensation for seepage and evaporation
Y'	Area prepared at the end of land preparation and transplanting time step
$Y_{ep'}$	Factor indicating the cumulative amount (evaporation and percolation)
Y_{ep}	Proportion of pre-saturated area
γ	Unit weight (kg/m^3)
ω	Water content (%)
γd	Dry unit weight (kg/m^3)
γsat	Saturated unit weight (kg/m^3)
ωsat	A saturated soil (%)

Annex C. Surface runoff analysis as inflow to the Telaga Tunjung Reservoir

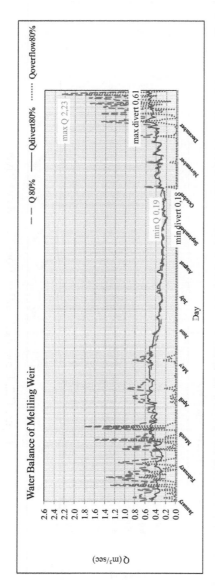

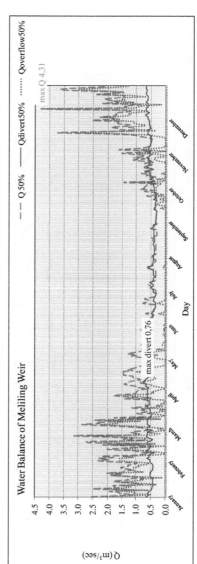

Figure C.1. Dependable inflow from overflow of Meliling Weir upstream of Telaga Tunjung Reservoir

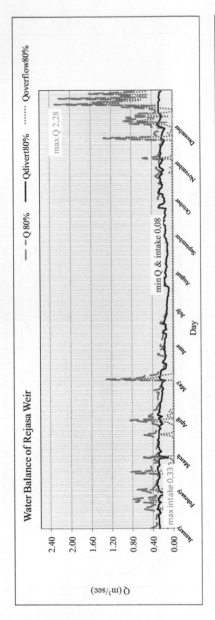

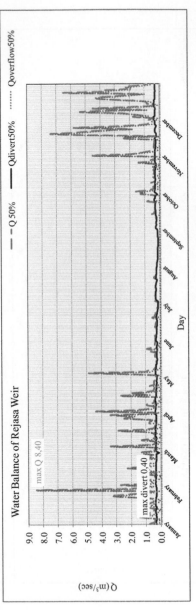

Figure C.2. Dependable inflow from Yeh Mawa River

Annex D. Analysis of rainfall data

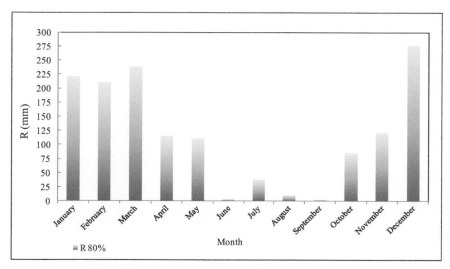

Figure D.1. Graph of monthly dependable rainfall

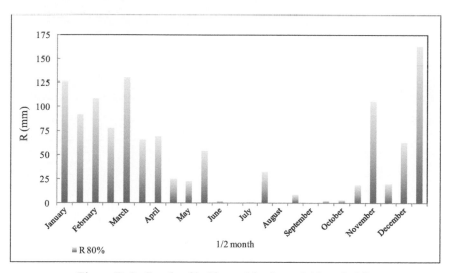

Figure D.2. Graph of half monthly dependable rainfall

Annex E. Reference evapotranspiration

Table E.1. Reference evapotranspiration

Month		T	RH (%)	n/N (%)	u (m/s)	ea (mbar)	w	(1 - w)	f(t)	ed (mbar)	ed - ea (mbar)	Ra (mbar)	Rs (mbar)	f (ed) (mbar)	f(n/N)	f(u)	Rn 1 (mm/day)	c	Eto* (mm/day)	Ep (mm/day)
January	1	25.3	85.2	50.0	1.3	32.4	0.7	0.3	15.7	27.6	4.8	16.1	8.4	0.1	0.6	0.6	0.9	1.04	4.7	4.9
	2	25.5	85.4	19.7	0.8	32.7	0.7	0.3	15.8	27.9	4.8	16.1	5.7	0.1	0.3	0.5	0.5	1.04	3.4	3.6
	3	24.9	85.4	44.8	1.1	31.6	0.7	0.3	15.6	27.0	4.6	16.1	7.9	0.1	0.5	0.5	0.9	1.04	4.4	4.6
	4	24.7	83.4	40.0	1.2	31.1	0.7	0.3	15.6	26.0	5.2	16.1	7.5	0.1	0.5	0.6	0.8	1.04	4.3	4.5
	5	25.3	84.4	36.4	0.7	32.3	0.7	0.3	15.7	27.2	5.0	16.1	7.2	0.1	0.4	0.4	0.7	1.04	4.0	4.2
	6	25.1	87.0	52.1	1.0	32.0	0.7	0.3	15.7	27.8	4.2	16.1	8.6	0.1	0.6	0.5	1.0	1.04	4.6	4.8
	7	24.6	88.0	17.3	1.1	30.9	0.7	0.3	15.6	27.2	3.7	16.1	5.5	0.1	0.3	0.5	0.4	1.04	3.2	3.4
	8	24.7	83.6	48.1	1.1	31.1	0.7	0.3	15.6	26.0	5.1	16.1	8.2	0.1	0.5	0.5	1.0	1.04	4.6	4.7
	9	24.2	86.4	61.4	1.3	30.1	0.7	0.3	15.5	26.0	4.1	16.1	9.4	0.1	0.7	0.6	1.2	1.04	4.9	5.1
	10	24.1	85.2	40.0	1.0	29.9	0.7	0.3	15.5	25.5	4.4	16.1	7.5	0.1	0.5	0.5	0.8	1.04	4.1	4.3
	11	24.6	84.8	41.6	1.3	31.0	0.7	0.3	15.6	26.3	4.7	16.1	7.6	0.1	0.5	0.6	0.8	1.04	4.3	4.5
	12	24.2	84.0	65.3	1.3	30.1	0.7	0.3	15.5	25.3	4.8	16.1	9.7	0.1	0.7	0.6	1.3	1.04	5.2	5.4
	13	24.2	79.8	35.8	1.9	30.1	0.7	0.3	15.5	24.0	6.1	16.1	7.1	0.1	0.4	0.7	0.8	1.04	4.5	4.7
	14	25.0	84.0	54.7	1.4	31.6	0.7	0.3	15.6	26.6	5.1	16.1	8.8	0.1	0.6	0.6	1.0	1.04	4.9	5.1
	15	24.8	89.0	49.2	1.1	31.2	0.7	0.3	15.6	27.8	3.4	16.1	8.3	0.1	0.5	0.5	0.9	1.04	4.4	4.6
	16	24.7	83.2	34.5	2.3	31.1	0.7	0.3	15.6	25.9	5.2	16.1	7.0	0.1	0.4	0.8	0.7	1.04	4.4	4.6
	17	24.8	83.0	27.2	1.7	31.3	0.7	0.3	15.6	26.0	5.3	16.1	6.4	0.1	0.3	0.7	0.6	1.04	4.0	4.2
	18	25.4	81.2	27.2	1.8	32.4	0.7	0.3	15.7	26.3	6.1	16.1	6.4	0.1	0.3	0.7	0.6	1.04	4.2	4.3
	19	25.1	81.0	9.3	1.2	31.9	0.7	0.3	15.7	25.9	6.1	16.1	4.8	0.1	0.2	0.5	0.3	1.04	3.3	3.4
	20	25.2	81.6	33.0	1.1	32.1	0.7	0.3	15.7	26.2	5.9	16.1	6.9	0.1	0.4	0.5	0.7	1.04	4.1	4.3
	21	25.6	85.6	37.3	1.1	32.9	0.8	0.2	15.8	28.2	4.7	16.1	7.3	0.1	0.4	0.5	0.7	1.04	4.2	4.3
	22	24.8	84.4	50.1	1.0	31.3	0.7	0.3	15.6	26.4	4.9	16.1	8.4	0.1	0.6	0.4	1.0	1.04	4.6	4.7
	23	25.2	87.0	32.7	0.7	32.2	0.7	0.3	15.7	28.0	4.2	16.1	6.9	0.1	0.4	0.4	0.7	1.04	3.8	4.0
	24	25.4	87.8	40.6	1.0	32.4	0.7	0.3	15.7	28.5	4.0	16.1	7.6	0.1	0.5	0.5	0.8	1.04	4.2	4.3
	25	24.8	82.8	53.0	1.0	31.4	0.7	0.3	15.6	26.0	5.4	16.1	8.6	0.1	0.6	0.5	1.0	1.04	4.7	4.9
	26	25.4	82.4	42.0	1.1	32.6	0.7	0.3	15.7	26.9	5.7	16.1	7.7	0.1	0.5	0.5	0.8	1.04	4.4	4.6
	27	25.3	88.2	29.4	1.2	32.3	0.7	0.3	15.7	28.5	3.8	16.1	6.6	0.1	0.4	0.6	0.6	1.04	3.8	3.9
	28	25.8	88.2	45.9	0.7	33.4	0.8	0.2	15.8	29.4	3.9	16.1	8.0	0.1	0.5	0.4	0.8	1.04	4.3	4.5
	29	25.4	84.8	39.4	1.4	32.5	0.7	0.3	15.7	27.5	4.9	16.1	7.5	0.1	0.5	0.6	0.8	1.04	4.3	4.5
	30	25.2	87.6	12.8	0.9	32.1	0.7	0.3	15.7	28.1	4.0	16.1	5.1	0.1	0.2	0.5	0.4	1.04	3.1	3.2
	31	25.2	82.0	30.8	1.1	32.1	0.7	0.3	15.7	26.3	5.8	16.1	6.7	0.1	0.4	0.5	0.7	1.04	4.0	4.2

Reservoir operation for water supply to Subak irrigation schemes in Yeh Ho River Basin

Table E.1. *continued*

	Month		T	RH (%)	n/N (%)	u (m/s)	ea (mbar)	w	(1 - w)	f(t)	cd (mbar)	cd - ea (mbar)	Ra (mbar)	Rs (mbar)	f (cd) (mbar)	f(n/N)	f(u)	Rn 1 (mm/day)	c	Eto* (mm/day)	Ep (mm/day)
2	February	1	24.7	90.0	17.1	0.9	31.2	0.7	0.3	15.6	28.1	3.1	16.1	5.5	0.1	0.3	0.5	0.4	1.05	3.1	3.3
		2	24.7	84.8	40.5	2.2	31.2	0.7	0.3	15.6	26.4	4.7	16.1	7.5	0.1	0.5	0.8	0.8	1.05	4.6	4.8
		3	23.9	92.6	15.6	0.8	29.6	0.7	0.3	15.4	27.4	2.2	16.1	5.4	0.1	0.2	0.5	0.4	1.05	2.9	3.1
		4	23.9	84.4	47.9	1.2	29.5	0.7	0.3	15.4	24.9	4.6	16.1	8.2	0.1	0.5	0.6	1.0	1.05	4.5	4.7
		5	24.9	85.8	36.9	1.3	31.4	0.7	0.3	15.6	27.0	4.5	16.1	7.2	0.1	0.4	0.6	0.8	1.05	4.1	4.3
		6	25.0	84.0	48.5	1.8	31.7	0.7	0.3	15.6	26.6	5.1	16.1	8.2	0.1	0.5	0.7	0.9	1.05	4.8	5.0
		7	25.7	89.2	52.5	1.8	33.2	0.8	0.2	15.8	29.6	3.6	16.1	8.6	0.1	0.6	0.7	0.9	1.05	4.8	5.0
		8	26.1	85.2	38.3	2.0	34.0	0.8	0.2	15.9	28.9	5.0	16.1	7.3	0.1	0.4	0.7	0.7	1.05	4.5	4.7
		9	24.7	80.2	39.3	1.4	31.1	0.7	0.3	15.6	24.9	6.2	16.1	7.4	0.1	0.5	0.6	0.8	1.05	4.5	4.7
		10	25.8	81.0	67.8	1.0	33.3	0.8	0.2	15.8	26.9	6.3	16.1	9.9	0.1	0.7	0.5	1.3	1.05	5.4	5.7
		11	25.6	81.0	55.7	1.3	32.9	0.8	0.2	15.8	26.6	6.2	16.1	8.9	0.1	0.6	0.6	1.1	1.05	5.1	5.3
		12	25.4	80.6	35.7	1.2	32.4	0.7	0.3	15.7	26.1	6.3	16.1	7.1	0.1	0.4	0.6	0.8	1.05	4.3	4.5
		13	25.6	87.8	34.0	1.3	32.8	0.8	0.2	15.8	28.8	4.0	16.1	7.0	0.1	0.4	0.6	0.7	1.05	4.0	4.2
		14	25.8	87.0	27.1	1.2	33.4	0.8	0.2	15.8	29.1	4.3	16.1	6.4	0.1	0.3	0.6	0.6	1.05	3.8	4.0
		15	26.1	88.2	16.7	0.8	34.0	0.8	0.2	15.9	30.0	4.0	16.1	5.5	0.1	0.3	0.5	0.4	1.05	3.3	3.4
		16	24.8	86.0	26.4	0.9	31.3	0.7	0.3	15.6	26.9	4.4	16.1	6.3	0.1	0.3	0.5	0.6	1.05	3.6	3.8
		17	25.8	82.6	15.3	0.8	33.2	0.8	0.2	15.8	27.5	5.8	16.1	5.4	0.1	0.2	0.4	0.4	1.05	3.4	3.5
		18	25.7	83.2	14.5	0.9	33.2	0.8	0.2	15.8	27.6	5.6	16.1	5.3	0.1	0.2	0.5	0.4	1.05	3.3	3.5
		19	25.9	83.2	29.3	0.9	33.4	0.8	0.2	15.8	27.8	5.6	16.1	6.6	0.1	0.4	0.5	0.6	1.05	3.9	4.1
		20	25.7	87.6	38.9	1.0	33.2	0.8	0.2	15.8	29.1	4.1	16.1	7.4	0.1	0.5	0.5	0.7	1.05	4.1	4.4
		21	25.8	88.0	27.3	1.0	33.4	0.8	0.2	15.8	29.4	4.0	16.1	6.4	0.1	0.3	0.5	0.6	1.05	3.7	3.9
		22	25.6	83.0	47.8	0.9	32.8	0.8	0.2	15.8	27.3	5.6	16.1	8.2	0.1	0.5	0.5	0.9	1.05	4.6	4.8
		23	25.4	83.0	14.0	0.8	32.5	0.7	0.3	15.7	27.0	5.5	16.1	5.2	0.1	0.2	0.5	0.4	1.05	3.3	3.4
		24	25.6	84.0	36.8	0.9	32.9	0.8	0.2	15.8	27.6	5.3	16.1	7.2	0.1	0.4	0.5	0.7	1.05	4.1	4.4
		25	25.7	88.0	63.8	2.2	33.2	0.8	0.2	15.8	29.2	4.0	16.1	9.6	0.1	0.7	0.8	1.1	1.05	5.1	5.3
		26	25.6	89.0	37.5	1.0	33.0	0.8	0.2	15.8	29.4	3.6	16.1	7.3	0.1	0.4	0.5	0.7	1.05	4.3	4.5
		27	25.6	91.8	33.5	1.0	32.9	0.8	0.2	15.8	30.2	2.7	16.1	6.9	0.1	0.4	0.5	0.6	1.05	3.8	4.0
		28	25.8	87.6	38.4	1.1	33.3	0.8	0.2	15.8	29.2	4.1	16.1	7.4	0.1	0.4	0.5	0.7	1.05	4.1	4.3

Table E.1. continued

Month		T	RH (%)	n/N (%)	u (m/s)	ea (mbar)	w	(1-w)	f(t)	ed (mbar)	ed-ea (mbar)	Ra (mbar)	Rs (mbar)	f(ed) (mbar)	f(n/N)	f(u)	Rn 1 (mm/day)	c	Eto* (mm/day)	Ep (mm/day)
3 March	1	25.5	86.3	40.7	1.1	32.7	0.7	0.3	15.8	28.2	4.5	15.5	7.3	0.1	0.5	0.5	0.8	1.06	4.1	4.3
	2	25.5	83.3	41.3	1.6	32.7	0.7	0.3	15.8	27.2	5.5	15.5	7.3	0.1	0.5	0.6	0.8	1.06	4.4	4.6
	3	25.3	80.8	40.3	2.1	32.4	0.7	0.3	15.7	26.2	6.2	15.5	7.2	0.1	0.5	0.8	0.8	1.06	4.6	4.9
	4	24.8	88.3	39.8	1.3	31.3	0.7	0.3	15.6	27.6	3.7	15.5	7.2	0.1	0.5	0.6	0.8	1.06	4.0	4.2
	5	24.9	80.0	36.6	1.0	31.5	0.7	0.3	15.6	25.2	6.3	15.5	6.9	0.1	0.4	0.5	0.8	1.06	4.1	4.3
	6	24.2	86.8	37.9	1.3	30.2	0.7	0.3	15.5	26.2	4.0	15.5	7.0	0.1	0.4	0.6	0.8	1.06	3.9	4.2
	7	24.6	84.3	38.7	2.1	30.9	0.7	0.3	15.6	26.0	4.9	15.5	7.1	0.1	0.4	0.8	0.8	1.06	4.3	4.6
	8	24.1	84.0	37.8	2.3	30.0	0.7	0.3	15.5	25.2	4.8	15.5	7.0	0.1	0.4	0.8	0.8	1.06	4.3	4.5
	9	24.8	81.8	38.0	2.2	31.2	0.7	0.3	15.6	25.5	5.7	15.5	7.0	0.1	0.4	0.8	0.8	1.06	4.5	4.7
	10	25.1	89.8	38.2	1.9	31.8	0.7	0.3	15.7	28.6	3.3	15.5	7.1	0.1	0.4	0.6	0.7	1.06	4.0	4.2
	11	25.3	88.0	38.5	1.3	32.4	0.7	0.3	15.7	28.5	3.9	15.5	7.1	0.1	0.4	0.4	0.7	1.06	4.0	4.2
	12	24.2	84.8	38.3	0.7	30.1	0.7	0.3	15.5	25.5	4.6	15.5	7.1	0.1	0.4	0.4	0.8	1.06	3.8	4.1
	13	24.6	88.8	40.6	0.8	30.9	0.7	0.3	15.6	27.4	3.5	15.5	7.3	0.1	0.5	0.5	0.8	1.06	3.9	4.1
	14	24.5	86.8	38.3	0.6	30.7	0.7	0.3	15.5	26.6	4.1	15.5	7.1	0.1	0.4	0.4	0.8	1.06	3.8	4.0
	15	24.0	86.8	39.0	0.5	29.8	0.7	0.3	15.4	25.8	3.9	15.5	7.1	0.1	0.5	0.4	0.8	1.06	3.7	4.0
	16	24.4	84.8	38.2	1.4	30.5	0.7	0.3	15.5	25.8	4.6	15.5	7.1	0.1	0.4	0.6	0.8	1.06	4.0	4.3
	17	23.4	87.5	38.9	1.0	28.6	0.7	0.3	15.3	25.0	3.6	15.5	7.1	0.1	0.5	0.5	0.8	1.06	3.8	4.0
	18	25.3	82.3	39.5	1.0	32.4	0.7	0.3	15.7	26.6	5.7	15.5	7.2	0.1	0.5	0.5	0.8	1.06	4.1	4.4
	19	24.0	82.7	39.8	0.9	29.8	0.7	0.3	15.4	24.6	5.2	15.5	7.2	0.1	0.5	0.5	0.9	1.06	4.0	4.2
	20	24.9	85.8	40.4	0.9	31.6	0.7	0.3	15.6	27.1	4.5	15.5	7.2	0.1	0.5	0.5	0.8	1.06	4.0	4.2
	21	24.3	81.7	39.3	0.3	30.4	0.7	0.3	15.5	24.8	5.6	15.5	7.1	0.1	0.4	0.3	0.8	1.06	3.8	4.0
	22	24.7	88.0	38.4	1.0	31.1	0.7	0.3	15.6	27.4	3.7	15.5	7.0	0.1	0.4	0.5	0.7	1.06	3.9	4.1
	23	24.5	90.0	37.3	1.1	30.6	0.7	0.3	15.5	27.6	3.1	15.5	7.1	0.1	0.4	0.5	0.8	1.06	3.8	4.0
	24	24.3	82.8	38.3	1.0	30.2	0.7	0.3	15.5	25.0	5.2	15.5	7.1	0.1	0.4	0.5	0.8	1.06	4.0	4.2
	25	25.3	86.3	39.2	0.8	32.3	0.7	0.3	15.7	27.9	4.4	15.5	7.1	0.1	0.5	0.5	0.8	1.06	3.9	4.2
	26	24.7	78.5	39.2	1.2	31.1	0.7	0.3	15.6	24.4	6.7	15.5	7.1	0.1	0.5	0.6	0.9	1.06	4.3	4.5
	27	23.7	89.0	37.9	1.1	29.1	0.7	0.3	15.4	25.9	3.2	15.5	7.0	0.1	0.4	0.5	0.8	1.06	3.7	4.0
	28	24.8	83.5	38.8	0.6	31.3	0.7	0.3	15.6	26.1	5.2	15.5	7.1	0.1	0.4	0.4	0.8	1.06	3.9	4.1
	29	24.5	84.5	39.1	0.6	30.6	0.7	0.3	15.5	25.9	4.8	15.5	7.1	0.1	0.5	0.4	0.8	1.06	3.9	4.1
	30	23.9	86.5	38.1	0.5	29.4	0.7	0.3	15.4	25.5	4.0	15.5	7.1	0.1	0.4	0.4	0.8	1.06	3.7	3.9
	31	23.7	92.0	38.3	0.5	29.2	0.7	0.3	15.4	26.9	2.3	15.5	7.1	0.1	0.4	0.4	0.8	1.06	3.6	3.8

Reservoir operation for water supply to Subak irrigation schemes in Yeh Ho River Basin

Table E.1. *continued*

Month		T	RH (%)	n/N (%)	u (m/s)	ea (mbar)	w	(1 - w)	f(t)	ed (mbar)	ed - ea (mbar)	Ra (mbar)	Rs (mbar)	f (ed) (mbar)	f(n/N)	f(u)	Rn 1 (mm/day)	c	Eto* (mm/day)	Ep (mm/day)
April	1	24.6	86.3	49.3	1.1	31.0	0.7	0.3	15.6	26.7	4.3	14.4	7.4	0.1	0.5	0.5	1.0	0.90	4.0	3.6
	2	24.0	87.0	33.9	0.4	29.7	0.7	0.3	15.6	25.9	3.9	14.4	6.2	0.1	0.4	0.4	0.7	0.90	3.3	2.9
	3	24.7	88.3	37.4	0.3	31.1	0.7	0.3	15.6	27.5	3.6	14.4	6.5	0.1	0.4	0.3	0.7	0.90	3.4	3.0
	4	24.5	89.7	44.3	0.5	30.6	0.7	0.3	15.5	27.5	3.2	14.4	7.0	0.1	0.5	0.4	0.8	0.90	3.6	3.2
	5	24.2	85.0	29.8	0.4	30.1	0.7	0.3	15.5	25.6	4.5	14.4	5.9	0.1	0.4	0.4	0.7	0.90	3.2	2.9
	6	24.0	87.3	31.9	0.9	29.8	0.7	0.3	15.4	26.0	3.8	14.4	6.1	0.1	0.4	0.5	0.7	0.90	3.3	3.0
	7	23.7	87.3	32.0	0.5	29.1	0.7	0.3	15.4	25.4	3.7	14.4	6.1	0.1	0.4	0.4	0.7	0.90	3.2	2.9
	8	24.8	91.0	42.9	0.6	31.3	0.7	0.3	15.6	28.5	2.8	14.4	6.9	0.1	0.5	0.4	0.8	0.90	3.6	3.2
	9	24.6	86.3	49.6	0.6	31.0	0.7	0.3	15.6	26.8	4.2	14.4	7.4	0.1	0.5	0.4	1.0	0.90	3.9	3.5
	10	24.3	87.0	54.2	0.6	30.3	0.7	0.3	15.5	26.3	3.9	14.4	7.8	0.1	0.6	0.4	1.0	0.90	4.0	3.6
	11	24.3	92.7	31.6	0.7	30.3	0.7	0.3	15.5	28.1	2.2	14.4	6.0	0.1	0.4	0.4	0.6	0.90	3.1	2.8
	12	24.7	86.0	36.7	0.5	31.0	0.7	0.3	15.6	26.7	4.3	14.4	6.4	0.1	0.4	0.4	0.8	0.90	3.4	3.1
	13	23.8	89.3	53.5	0.7	29.4	0.7	0.3	15.4	26.3	3.1	14.4	7.7	0.1	0.6	0.4	1.0	0.90	3.9	3.5
	14	24.4	86.7	41.1	0.7	30.6	0.7	0.3	15.5	26.5	4.1	14.4	6.8	0.1	0.5	0.4	0.8	0.90	3.6	3.2
	15	25.3	84.7	39.7	0.5	32.4	0.7	0.3	15.7	27.4	5.0	14.4	6.7	0.1	0.5	0.4	0.8	0.90	3.6	3.3
	16	25.2	85.0	59.0	0.7	32.1	0.7	0.3	15.7	27.3	4.8	14.4	8.2	0.1	0.6	0.4	1.1	0.90	4.3	3.9
	17	23.0	87.0	55.1	0.4	27.6	0.7	0.3	15.2	24.0	3.6	14.4	7.9	0.1	0.6	0.4	1.1	0.90	3.8	3.4
	18	23.7	84.3	58.2	0.8	29.1	0.7	0.3	15.4	24.5	4.6	14.4	8.1	0.1	0.6	0.5	1.2	0.90	4.1	3.7
	19	23.8	93.3	65.7	0.5	29.3	0.7	0.3	15.4	27.4	2.0	14.4	8.7	0.1	0.7	0.4	1.2	0.90	4.1	3.7
	20	24.2	83.7	47.3	0.4	30.1	0.7	0.3	15.5	25.2	4.9	14.4	7.3	0.1	0.5	0.4	1.0	0.90	3.8	3.4
	21	24.1	65.3	18.2	0.6	30.0	0.7	0.3	15.5	19.6	10.4	14.4	5.0	0.1	0.3	0.4	0.6	0.90	3.5	3.1
	22	24.8	89.7	63.0	0.7	31.2	0.7	0.3	15.6	28.0	3.2	14.4	8.5	0.1	0.7	0.4	1.1	0.90	4.3	3.8
	23	24.2	84.7	62.6	0.4	30.1	0.7	0.3	15.5	25.5	4.6	14.4	8.4	0.1	0.7	0.4	1.2	0.90	4.2	3.8
	24	24.6	85.7	35.4	0.4	30.9	0.7	0.3	15.6	26.5	4.4	14.4	6.3	0.1	0.4	0.4	0.7	0.90	3.4	3.0
	25	25.1	89.0	52.2	0.9	31.8	0.7	0.3	15.7	28.3	3.5	14.4	7.6	0.1	0.6	0.5	0.9	0.90	4.0	3.6
	26	24.1	77.0	34.7	0.4	29.9	0.7	0.3	15.5	23.0	6.9	14.4	6.3	0.1	0.4	0.4	0.8	0.90	3.5	3.2
	27	23.9	85.0	30.8	0.5	29.5	0.7	0.3	15.4	25.1	4.4	14.4	6.0	0.1	0.4	0.4	0.7	0.90	3.2	2.9
	28	24.3	86.3	32.1	0.6	30.4	0.7	0.3	15.5	26.2	4.2	14.4	6.1	0.1	0.4	0.4	0.7	0.90	3.3	3.0
	29	23.8	90.0	20.7	0.4	29.4	0.7	0.3	15.4	26.4	2.9	14.4	5.2	0.1	0.3	0.4	0.5	0.90	2.8	2.5
	30	24.1	87.0	29.1	0.3	29.8	0.7	0.3	15.4	26.0	3.9	14.4	5.8	0.1	0.4	0.3	0.6	0.90	3.1	2.8

Table E.1. *continued*

	Month		T	RH (%)	n/N (%)	u (m/s)	ea (mbar)	w	(1 - w)	f(t)	ed (mbar)	ed - ea (mbar)	Ra (mbar)	Rs (mbar)	f (ed) (mbar)	f(n/N)	f(u)	Rn 1 (mm/day)	c	Eto* (mm/day)	Ep (mm/day)
5	May	1	26.1	87.0	57.8	0.6	33.9	0.8	0.2	15.9	29.5	4.4	13.0	7.3	0.1	0.6	0.4	1.0	0.90	3.9	3.5
		2	25.6	89.5	28.1	0.7	33.0	0.8	0.2	15.8	29.5	3.5	13.0	5.2	0.1	0.4	0.4	0.6	0.90	2.9	2.6
		3	24.5	84.8	19.4	0.7	30.8	0.7	0.3	15.5	26.1	4.7	13.0	4.6	0.1	0.3	0.4	0.5	0.90	2.7	2.4
		4	24.8	87.5	42.3	0.6	31.4	0.7	0.3	15.6	27.5	3.9	13.0	6.2	0.1	0.5	0.4	0.8	0.90	3.3	3.0
		5	25.3	90.3	48.9	0.6	32.4	0.7	0.3	15.7	29.2	3.2	13.0	6.7	0.1	0.5	0.4	0.9	0.90	3.4	3.1
		6	24.9	87.3	59.6	0.7	31.5	0.7	0.3	15.6	27.5	4.0	13.0	7.5	0.1	0.6	0.4	1.1	0.90	3.8	3.4
		7	25.1	87.5	62.0	0.6	32.0	0.7	0.3	15.7	28.0	4.0	13.0	7.6	0.1	0.7	0.4	1.1	0.90	3.9	3.5
		8	24.8	87.5	28.8	0.6	31.3	0.7	0.3	15.6	27.4	3.9	13.0	5.3	0.1	0.4	0.4	0.6	0.90	2.9	2.6
		9	24.5	88.3	31.3	0.5	30.6	0.7	0.3	15.5	27.0	3.6	13.0	5.5	0.1	0.4	0.4	0.7	0.90	2.9	2.6
		10	35.3	85.3	60.0	0.6	52.4	0.8	0.2	17.9	44.7	7.7	13.0	7.5	0.0	0.6	0.4	0.5	0.90	4.8	4.3
		11	23.8	86.0	37.7	0.5	29.3	0.7	0.3	15.4	25.2	4.1	13.0	5.9	0.1	0.4	0.4	0.8	0.90	3.1	2.8
		12	24.4	81.0	47.8	0.5	30.6	0.7	0.3	15.5	24.8	5.8	13.0	6.6	0.1	0.5	0.4	1.0	0.90	3.5	3.2
		13	23.8	87.3	44.1	0.5	29.2	0.7	0.3	15.4	25.5	3.7	13.0	6.4	0.1	0.5	0.4	0.9	0.90	3.2	2.9
		14	24.0	84.3	53.5	0.6	29.8	0.7	0.3	15.4	25.1	4.7	13.0	7.0	0.1	0.6	0.4	1.1	0.90	3.6	3.2
		15	24.3	87.0	56.5	0.3	30.3	0.7	0.3	15.5	26.4	3.9	13.0	7.2	0.1	0.6	0.3	1.1	0.90	3.6	3.2
		16	24.1	79.3	65.7	0.7	30.0	0.7	0.3	15.5	23.7	6.2	13.0	7.9	0.1	0.7	0.4	1.3	0.90	4.1	3.7
		17	23.6	86.0	67.7	0.5	28.9	0.7	0.3	15.3	24.9	4.0	13.0	8.0	0.1	0.7	0.4	1.3	0.90	3.9	3.5
		18	23.3	81.8	60.0	0.5	28.4	0.7	0.3	15.3	23.2	5.2	13.0	7.5	0.1	0.6	0.4	1.3	0.90	3.7	3.3
		19	22.9	86.5	61.9	0.6	27.6	0.7	0.3	15.2	23.8	3.7	13.0	7.6	0.1	0.7	0.4	1.2	0.90	3.7	3.3
		20	24.0	80.7	46.6	0.3	29.6	0.7	0.3	15.4	23.9	5.7	13.0	6.5	0.1	0.5	0.3	1.0	0.90	3.4	3.1
		21	23.2	87.3	50.0	0.4	28.2	0.7	0.3	15.3	24.6	3.6	13.0	6.8	0.1	0.6	0.4	1.0	0.90	3.3	3.0
		22	24.3	79.0	53.3	0.7	30.3	0.7	0.3	15.5	23.9	6.4	13.0	7.0	0.1	0.6	0.4	1.1	0.90	3.8	3.4
		23	24.8	85.7	49.5	0.4	31.3	0.7	0.3	15.6	26.8	4.5	13.0	6.7	0.1	0.5	0.4	1.0	0.90	3.5	3.1
		24	24.6	88.7	59.1	0.7	30.9	0.7	0.3	15.6	27.4	3.5	13.0	7.4	0.1	0.6	0.4	1.1	0.90	3.7	3.3
		25	25.1	89.3	45.8	0.6	31.8	0.7	0.3	15.7	28.4	3.4	13.0	6.5	0.1	0.5	0.4	0.8	0.90	3.4	3.0
		26	24.5	86.7	55.8	0.5	30.6	0.7	0.3	15.5	26.6	4.1	13.0	7.2	0.1	0.6	0.4	1.1	0.90	3.6	3.3
		27	24.2	87.7	56.2	0.7	30.1	0.7	0.3	15.5	26.4	3.7	13.0	7.2	0.1	0.6	0.4	1.1	0.90	3.6	3.3
		28	23.3	85.3	61.9	0.7	28.3	0.7	0.3	15.3	24.1	4.2	13.0	7.6	0.1	0.7	0.4	1.2	0.90	3.8	3.4
		29	23.2	83.3	56.8	0.5	28.0	0.7	0.3	15.3	23.3	4.7	13.0	7.3	0.1	0.6	0.4	1.2	0.90	3.6	3.2
		30	23.6	88.8	60.3	0.6	28.9	0.7	0.3	15.3	25.6	3.2	13.0	7.5	0.1	0.6	0.4	1.2	0.90	3.6	3.3
		31	24.1	87.5	39.8	0.4	29.8	0.7	0.3	15.4	26.1	3.7	13.0	6.1	0.1	0.5	0.4	0.8	0.90	3.1	2.8

Reservoir operation for water supply to Subak irrigation schemes in Yeh Ho River Basin

Table E.1. *continued*

Month		T	RH (%)	n/N (%)	u (m/s)	ea (mbar)	w	(1 - w)	f(t)	ed (mbar)	ed - ea (mbar)	Ra (mbar)	Rs (mbar)	f (ed) (mbar)	f(n/N)	f(u)	Rn 1 (mm/day)	c	Eto* (mm/day)	Ep (mm/day)
6	June																			
	1	24.1	81.8	44.6	0.6	30.0	0.7	0.3	15.5	24.5	5.5	12.4	6.1	0.1	0.5	0.4	0.9	0.90	3.3	2.9
	2	24.6	83.5	45.0	0.4	30.9	0.7	0.3	15.6	25.8	5.1	12.4	6.1	0.1	0.5	0.4	0.9	0.90	3.2	2.9
	3	24.6	83.5	25.5	0.5	31.0	0.7	0.3	15.6	25.9	5.1	12.4	4.8	0.1	0.3	0.4	0.6	0.90	2.7	2.5
	4	24.7	91.0	50.7	0.5	31.2	0.7	0.3	15.6	28.4	2.8	12.4	6.5	0.1	0.6	0.4	0.9	0.90	3.2	2.9
	5	24.3	88.0	35.5	0.6	30.3	0.7	0.3	15.5	26.6	3.6	12.4	5.5	0.1	0.4	0.4	0.7	0.90	2.9	2.6
	6	25.1	88.5	40.3	0.6	32.0	0.7	0.3	15.7	28.3	3.7	12.4	5.8	0.1	0.5	0.4	0.8	0.90	3.1	2.7
	7	25.1	83.8	45.4	0.5	32.0	0.7	0.3	15.7	26.8	5.2	12.4	6.1	0.1	0.5	0.4	0.9	0.90	3.3	3.0
	8	24.7	84.0	42.3	0.6	31.0	0.7	0.3	15.6	26.1	5.0	12.4	5.9	0.1	0.5	0.4	0.9	0.90	3.2	2.9
	9	24.2	83.5	38.0	0.7	30.1	0.7	0.3	15.5	25.1	5.0	12.4	5.7	0.1	0.4	0.4	0.8	0.90	3.1	2.8
	10	24.3	92.0	31.8	0.6	30.4	0.7	0.3	15.5	27.9	2.4	12.4	5.2	0.1	0.4	0.4	0.6	0.90	2.7	2.4
	11	24.4	82.8	46.3	0.7	30.5	0.7	0.3	15.5	25.3	5.3	12.4	6.2	0.1	0.5	0.4	1.0	0.90	3.3	3.0
	12	24.0	82.0	43.0	0.6	29.7	0.7	0.3	15.4	24.3	5.3	12.4	6.0	0.1	0.5	0.4	0.9	0.90	3.2	2.9
	13	24.3	77.3	36.3	0.8	30.3	0.7	0.3	15.5	23.4	6.9	12.4	5.5	0.1	0.4	0.5	0.8	0.90	3.3	2.9
	14	24.0	76.0	36.7	0.7	29.6	0.7	0.3	15.4	22.5	7.1	12.4	5.6	0.1	0.4	0.4	0.9	0.90	3.2	2.9
	15	23.9	79.0	57.1	0.9	29.5	0.7	0.3	15.4	23.3	6.2	12.4	6.9	0.1	0.6	0.5	1.2	0.90	3.7	3.3
	16	23.8	88.8	47.9	0.5	29.3	0.7	0.3	15.4	26.0	3.3	12.4	6.3	0.1	0.5	0.4	0.9	0.90	3.1	2.8
	17	24.9	84.8	34.1	0.6	31.4	0.7	0.3	15.6	26.6	4.8	12.4	5.4	0.1	0.4	0.4	0.7	0.90	3.0	2.7
	18	23.6	86.0	31.8	0.5	28.9	0.7	0.3	15.3	24.8	4.0	12.4	5.2	0.1	0.4	0.4	0.7	0.90	2.8	2.5
	19	24.4	88.3	7.0	0.6	30.6	0.7	0.3	15.5	27.0	3.6	12.4	3.6	0.1	0.2	0.4	0.3	0.90	2.2	*1.9*
	20	24.5	89.3	3.9	0.5	30.7	0.7	0.3	15.5	27.4	3.3	12.4	3.4	0.1	0.1	0.4	0.2	0.90	2.0	*1.8*
	21	24.0	84.3	26.0	0.6	29.8	0.7	0.3	15.4	25.1	4.7	12.4	4.8	0.1	0.3	0.4	0.6	0.90	2.7	2.5
	22	23.8	93.0	44.3	0.5	29.4	0.7	0.3	15.4	27.3	2.1	12.4	6.1	0.1	0.5	0.4	0.8	0.90	2.9	2.6
	23	22.8	86.7	41.7	0.6	27.2	0.7	0.3	15.2	23.6	3.6	12.4	5.9	0.1	0.5	0.4	0.9	0.90	3.0	2.7
	24	24.1	83.7	53.4	0.8	29.9	0.7	0.3	15.5	25.0	4.9	12.4	6.7	0.1	0.6	0.5	1.1	0.90	3.5	3.1
	25	24.0	89.7	27.5	0.4	29.6	0.7	0.3	15.4	26.6	3.1	12.4	5.0	0.1	0.3	0.4	0.6	0.90	2.6	2.3
	26	24.1	88.0	31.0	0.6	29.9	0.7	0.3	15.5	26.3	3.6	12.4	5.2	0.1	0.4	0.4	0.7	0.90	2.8	2.5
	27	23.7	86.0	29.5	0.5	29.1	0.7	0.3	15.4	25.0	4.1	12.4	5.1	0.1	0.4	0.4	0.7	0.90	2.7	2.4
	28	23.8	84.0	39.6	0.5	29.3	0.7	0.3	15.4	24.6	4.7	12.4	5.8	0.1	0.5	0.4	0.9	0.90	3.0	2.7
	29	23.5	83.0	35.7	0.4	28.7	0.7	0.3	15.3	23.8	4.9	12.4	5.5	0.1	0.4	0.4	0.8	0.90	2.9	2.6
	30	24.0	84.7	68.2	0.8	29.7	0.7	0.3	15.4	25.2	4.6	12.4	7.7	0.1	0.7	0.5	1.3	0.90	3.8	3.4

Table E.1. *continued*

	Month		T	RH (%)	n/N (%)	u (m/s)	ea (mbar)	w	(1 - w)	f(t)	ed (mbar)	ed - ea (mbar)	Ra (mbar)	Rs (mbar)	f (ed) (mbar)	f(n/N)	f(u)	Rn 1 (mm/day)	c	Eto* (mm/day)	Ep (mm/day)
7	July	1	24.2	81.3	35.4	0.6	30.0	0.7	0.3	15.5	24.4	5.6	12.6	5.6	0.1	0.4	0.4	0.8	0.90	3.1	2.8
		2	23.7	77.0	30.5	0.6	29.1	0.7	0.3	15.4	22.4	6.7	12.6	5.2	0.1	0.4	0.4	0.8	0.90	3.0	2.7
		3	24.2	87.0	43.0	0.7	30.0	0.7	0.3	15.5	26.1	3.9	12.6	6.1	0.1	0.5	0.4	0.9	0.90	3.2	2.8
		4	24.3	82.7	47.1	0.6	30.2	0.7	0.3	15.5	25.0	5.2	12.6	6.4	0.1	0.5	0.4	1.0	0.90	3.4	3.0
		5	24.0	87.3	55.6	1.2	29.7	0.7	0.3	15.4	25.9	3.8	12.6	6.9	0.1	0.6	0.6	1.1	0.90	3.6	3.2
		6	23.7	86.0	35.9	0.5	29.0	0.7	0.3	15.4	25.0	4.1	12.6	5.6	0.1	0.4	0.4	0.8	0.90	2.9	2.6
		7	22.9	85.3	34.1	0.6	27.5	0.7	0.3	15.2	23.4	4.1	12.6	5.5	0.1	0.4	0.4	0.8	0.90	2.9	2.6
		8	23.0	82.3	52.8	0.7	27.6	0.7	0.3	15.2	22.7	4.9	12.6	6.7	0.1	0.6	0.4	1.1	0.90	3.4	3.1
		9	23.7	85.3	34.9	0.7	29.0	0.7	0.3	15.4	24.8	4.3	12.6	5.5	0.1	0.4	0.4	0.8	0.90	3.0	2.7
		10	22.6	89.5	42.2	0.7	26.9	0.7	0.3	15.1	24.0	2.8	12.6	6.0	0.1	0.5	0.4	0.9	0.90	2.9	2.7
		11	22.9	85.8	53.1	0.8	27.4	0.7	0.3	15.2	23.5	3.9	12.6	6.8	0.1	0.6	0.5	1.1	0.90	3.4	3.0
		12	22.8	89.0	45.3	0.6	27.4	0.7	0.5	15.2	24.3	3.0	12.6	6.2	0.1	0.5	0.4	0.9	0.90	3.0	2.7
		13	22.9	88.8	42.6	0.8	27.4	0.7	0.5	15.2	24.3	3.1	12.6	6.1	0.1	0.5	0.5	0.9	0.90	3.0	2.7
		14	23.7	84.8	29.8	0.6	29.0	0.7	0.5	15.4	24.6	4.4	12.6	5.2	0.1	0.4	0.4	0.7	0.90	2.8	2.5
		15	23.2	92.5	23.8	0.5	28.0	0.7	0.3	15.3	25.9	2.1	12.6	4.8	0.1	0.3	0.4	0.6	0.90	2.4	2.2
		16	23.2	88.0	42.3	0.6	28.0	0.7	0.3	15.3	24.7	3.4	12.6	6.0	0.1	0.5	0.4	0.9	0.90	3.0	2.7
		17	22.6	84.3	39.4	0.7	26.8	0.7	0.3	15.1	22.6	4.2	12.6	5.8	0.1	0.5	0.4	0.9	0.90	3.0	2.7
		18	22.9	83.8	38.9	0.4	27.4	0.7	0.3	15.2	23.0	4.5	12.6	5.8	0.1	0.5	0.4	0.9	0.90	3.0	2.7
		19	22.7	76.5	39.3	0.6	27.1	0.7	0.3	15.2	20.7	6.4	12.6	5.8	0.1	0.5	0.4	1.0	0.90	3.2	2.9
		20	22.8	86.0	33.5	0.6	27.3	0.7	0.3	15.2	23.5	3.8	12.6	5.4	0.1	0.4	0.4	0.8	0.90	2.8	2.5
		21	23.1	82.3	27.8	0.6	27.9	0.7	0.3	15.2	23.0	5.0	12.6	5.0	0.1	0.3	0.4	0.7	0.90	2.8	2.5
		22	23.2	87.5	30.7	0.6	28.1	0.7	0.3	15.3	24.6	3.5	12.6	5.2	0.1	0.4	0.4	0.7	0.90	2.7	2.5
		23	23.3	84.8	27.0	0.7	28.4	0.7	0.3	15.3	24.0	4.3	12.6	5.0	0.1	0.3	0.4	0.7	0.90	2.7	2.5
		24	22.7	87.0	35.1	0.7	27.1	0.7	0.3	15.2	23.6	3.5	12.6	5.5	0.1	0.4	0.4	0.8	0.90	2.9	2.6
		25	22.1	90.3	29.5	1.0	25.9	0.7	0.3	15.0	23.4	2.5	12.6	5.2	0.1	0.4	0.5	0.7	0.90	2.6	2.4
		26	22.7	85.5	22.9	0.5	27.2	0.7	0.3	15.2	23.2	3.9	12.6	4.7	0.1	0.3	0.4	0.6	0.90	2.5	2.3
		27	22.5	88.8	35.5	0.7	26.6	0.7	0.3	15.1	23.6	3.0	12.6	5.6	0.1	0.4	0.4	0.8	0.90	2.8	2.5
		28	21.7	84.5	42.1	0.4	25.1	0.7	0.3	14.9	21.2	3.9	12.6	6.0	0.1	0.5	0.4	1.0	0.90	2.9	2.6
		29	23.5	88.8	43.1	0.6	28.6	0.7	0.3	15.3	25.4	3.2	12.6	6.1	0.1	0.5	0.4	0.9	0.90	3.0	2.7
		30	24.1	85.8	59.9	0.9	29.9	0.7	0.3	15.5	25.6	4.3	12.6	7.2	0.1	0.6	0.5	1.2	0.90	3.7	3.3
		31	21.9	85.3	32.2	1.0	25.5	0.7	0.3	15.0	21.8	3.7	12.6	5.3	0.1	0.4	0.5	0.8	0.90	2.8	2.5

Reservoir operation for water supply to Subak irrigation schemes in Yeh Ho River Basin

Table E.1. *continued*

Month		T	RH (%)	n/N (%)	u (m/s)	ea (mbar)	w	(1-w)	f(t)	ed (mbar)	ed-ea (mbar)	Ra (mbar)	Rs (mbar)	f(cd) (mbar)	f(n/N)	f(u)	Rn 1 (mm/day)	c	Eto* (mm/day)	Ep (mm/day)
August	1	24.6	92.0	50.3	0.9	31.0	0.7	0.3	15.6	28.5	2.5	13.5	7.1	0.1	0.6	0.5	0.9	1.00	3.6	3.6
	2	25.8	84.3	49.0	0.8	33.2	0.8	0.2	15.8	28.0	5.2	13.5	7.0	0.1	0.5	0.5	0.9	1.00	3.8	3.8
	3	25.2	84.0	48.2	0.6	32.1	0.7	0.3	15.7	27.0	5.1	13.5	6.9	0.1	0.5	0.4	0.9	1.00	3.7	3.7
	4	25.2	89.5	53.4	0.6	32.2	0.7	0.3	15.7	28.8	3.4	13.5	7.3	0.1	0.6	0.4	0.9	1.00	3.7	3.7
	5	25.1	84.5	41.8	0.8	32.0	0.7	0.3	15.7	27.0	5.0	13.5	6.4	0.1	0.5	0.4	0.8	1.00	3.5	3.5
	6	23.7	77.5	31.2	0.6	29.1	0.7	0.3	15.4	22.5	6.5	13.5	5.7	0.1	0.4	0.4	0.8	1.00	3.3	3.3
	7	23.5	84.5	42.2	0.7	28.7	0.7	0.3	15.3	24.2	4.4	13.5	6.5	0.1	0.5	0.4	0.9	1.00	3.4	3.4
	8	23.3	85.5	37.1	0.8	28.2	0.7	0.3	15.3	24.1	4.1	13.5	6.1	0.1	0.4	0.4	0.8	1.00	3.2	3.2
	9	22.5	89.5	36.6	0.5	26.6	0.7	0.3	15.1	23.8	2.8	13.5	6.1	0.1	0.4	0.4	0.8	1.00	3.0	3.0
	10	23.2	89.8	44.5	0.6	28.2	0.7	0.3	15.3	25.3	2.9	13.5	6.6	0.1	0.5	0.4	0.9	1.00	3.3	3.3
	11	22.6	82.0	47.6	0.7	26.8	0.7	0.3	15.1	22.0	4.8	13.5	6.9	0.1	0.6	0.4	1.1	1.00	3.6	3.6
	12	22.9	85.3	51.4	0.6	27.5	0.7	0.3	15.2	23.4	4.1	13.5	7.1	0.1	0.6	0.4	1.1	1.00	3.2	3.2
	13	21.9	90.3	46.0	0.7	25.5	0.7	0.3	15.0	23.0	2.5	13.5	6.7	0.1	0.5	0.5	1.0	1.00	4.5	4.5
	14	21.8	62.3	62.9	1.1	25.3	0.7	0.3	15.0	15.8	9.6	13.5	8.0	0.2	0.7	0.5	1.6	1.00	4.0	4.0
	15	21.2	63.5	52.8	0.7	24.1	0.7	0.3	14.8	15.3	8.8	13.5	7.2	0.2	0.6	0.5	1.4	1.00	4.4	4.4
	16	21.2	62.8	64.1	0.9	24.1	0.7	0.3	14.8	15.1	9.0	13.5	8.1	0.2	0.7	0.5	1.7	1.00	4.2	4.2
	17	21.5	64.3	62.8	0.6	24.7	0.7	0.3	14.9	15.9	8.8	13.5	8.0	0.2	0.7	0.4	1.6	1.00	3.8	3.8
	18	21.5	88.8	61.4	1.8	24.6	0.7	0.3	14.9	21.8	2.8	13.5	7.9	0.1	0.7	0.7	1.3	1.00	3.4	3.4
	19	21.3	84.5	49.7	0.6	24.4	0.7	0.3	14.9	20.6	3.8	13.5	7.0	0.1	0.5	0.4	1.1	1.00	3.8	3.8
	20	24.7	84.8	51.6	1.0	31.2	0.7	0.3	15.6	26.4	4.8	13.5	7.2	0.1	0.6	0.5	1.0	1.00	3.5	3.5
	21	24.1	86.5	48.8	0.6	29.8	0.7	0.3	15.4	25.8	4.0	13.5	7.0	0.1	0.5	0.4	1.1	1.00	3.4	3.4
	22	22.0	84.5	49.1	0.7	25.7	0.7	0.3	15.0	21.7	4.0	13.5	7.0	0.1	0.5	0.4	0.7	1.00	2.9	2.9
	23	22.1	81.0	26.4	0.6	25.8	0.7	0.3	15.0	20.9	4.9	13.5	5.3	0.1	0.3	0.4	0.7	1.00	2.6	2.6
	24	20.2	86.8	25.4	0.7	22.0	0.7	0.3	14.6	19.1	2.9	13.5	5.2	0.1	0.3	0.4	0.8	1.00	2.7	2.7
	25	20.3	86.5	29.2	0.4	22.2	0.7	0.3	14.6	19.2	3.0	13.5	5.5	0.1	0.4	0.4	0.9	1.00	2.7	2.7
	26	19.2	87.0	34.1	0.5	20.0	0.7	0.3	14.4	17.4	2.6	13.5	5.9	0.2	0.4	0.4	1.1	1.00	3.4	3.4
	27	20.5	71.3	41.5	0.6	22.6	0.7	0.3	14.7	16.1	6.5	13.5	6.4	0.2	0.5	0.4	1.0	1.00	3.4	3.4
	28	22.8	83.7	45.3	0.5	27.3	0.7	0.3	15.2	22.8	4.5	13.5	6.7	0.1	0.5	0.5	0.8	1.00	3.1	3.1
	29	21.9	82.7	33.7	0.9	25.4	0.7	0.3	15.0	21.0	4.4	13.5	5.8	0.1	0.4	0.5	0.8	1.00	3.1	3.1
	30	22.6	84.3	64.6	0.8	26.8	0.7	0.3	15.1	22.6	4.2	13.5	8.1	0.1	0.7	0.5	1.3	1.00	4.0	4.0
	31	22.8	88.3	58.9	0.8	27.3	0.7	0.3	15.2	24.1	3.2	13.5	7.7	0.1	0.6	0.5	1.2	1.00	3.7	3.7

8

Table E.1. *continued*

Month		T	RH (%)	n/N (%)	u (m/s)	ea (mber)	w	(1 - w)	f(t)	ed (mbar)	ed - ea (mbar)	Ra (mbar)	Rs (mbar)	f (ed) (mbar)	f(n/N)	f(u)	Rn 1 (mm/day)	c	Eto* (mm/day)	Ep (mm/day)
9	September	1	24.9	86.8	18.8	0.9	31.5	0.7	0.3	15.6	27.3	4.2	14.7	5.2	0.1	0.3	0.5	0.5	1.10	3.0	3.4
		2	25.0	88.5	19.3	0.8	31.6	0.7	0.3	15.6	28.0	3.6	14.7	5.2	0.1	0.3	0.4	0.5	1.10	3.0	3.3
		3	23.9	91.0	32.2	0.3	29.6	0.7	0.3	15.4	26.9	2.7	14.7	6.2	0.1	0.4	0.3	0.7	1.10	3.2	3.5
		4	24.2	90.0	28.6	0.6	30.1	0.7	0.3	15.5	27.1	3.0	14.7	6.0	0.1	0.4	0.4	0.6	1.10	3.2	3.5
		5	23.6	90.5	32.7	0.8	28.9	0.7	0.3	15.3	26.2	2.7	14.7	6.3	0.1	0.4	0.5	0.7	1.10	3.3	3.6
		6	23.7	83.3	39.8	0.5	29.0	0.7	0.3	15.4	24.2	4.9	14.7	6.8	0.1	0.5	0.4	0.9	1.10	3.6	4.0
		7	24.2	88.8	19.8	0.8	30.1	0.7	0.3	15.5	26.8	3.4	14.7	5.3	0.1	0.3	0.4	0.5	1.10	2.9	3.2
		8	23.7	87.0	24.9	0.8	29.2	0.7	0.3	15.4	25.4	3.8	14.7	5.7	0.1	0.3	0.4	0.6	1.10	3.1	3.4
		9	24.7	83.0	27.3	0.5	31.1	0.7	0.3	15.6	25.8	5.3	14.7	5.8	0.1	0.3	0.4	0.6	1.10	3.3	3.6
		10	24.4	81.5	16.4	0.7	30.4	0.7	0.3	15.5	24.8	5.6	14.7	5.0	0.1	0.2	0.4	0.5	1.10	3.1	3.4
		11	23.2	86.8	25.8	0.5	28.1	0.7	0.3	15.3	24.4	3.7	14.7	5.7	0.1	0.3	0.4	0.6	1.10	3.1	3.4
		12	22.9	85.8	28.8	0.8	27.5	0.7	0.3	15.2	23.6	3.9	14.7	6.0	0.1	0.4	0.5	0.7	1.10	3.2	3.6
		13	23.9	85.5	47.0	0.9	29.6	0.7	0.3	15.4	25.3	4.3	14.7	7.4	0.1	0.5	0.5	1.0	1.10	3.9	4.3
		14	23.2	85.5	47.0	0.7	28.1	0.7	0.3	15.3	24.0	4.1	14.7	7.4	0.1	0.5	0.4	1.0	1.10	3.8	4.2
		15	34.7	88.5	51.8	0.7	51.2	0.8	0.2	17.8	45.3	5.9	14.7	7.8	0.0	0.6	0.4	0.4	1.10	5.0	5.5
		16	22.7	84.5	39.4	0.8	27.0	0.7	0.3	15.1	22.8	4.2	14.7	6.8	0.1	0.5	0.4	0.9	1.10	3.6	3.9
		17	25.0	89.5	34.6	0.7	31.8	0.7	0.3	15.7	28.4	3.3	14.7	6.4	0.1	0.4	0.4	0.7	1.10	3.5	3.8
		18	24.1	85.8	58.0	0.8	30.0	0.7	0.3	15.5	25.7	4.3	14.7	8.3	0.1	0.6	0.5	1.1	1.10	4.3	4.7
		19	24.7	87.3	59.8	0.9	31.1	0.7	0.3	15.6	27.1	4.0	14.7	8.4	0.1	0.6	0.5	1.1	1.10	4.4	4.8
		20	24.5	87.3	55.0	0.8	30.7	0.7	0.3	15.5	26.8	3.9	14.7	8.1	0.1	0.6	0.5	1.0	1.10	4.2	4.6
		21	24.7	88.0	51.9	0.8	31.1	0.7	0.3	15.6	27.3	3.7	14.7	7.8	0.1	0.6	0.5	1.0	1.10	4.1	4.5
		22	24.6	85.8	46.1	0.7	30.9	0.7	0.3	15.6	26.5	4.4	14.7	7.3	0.1	0.5	0.4	0.9	1.10	3.9	4.3
		23	24.7	89.3	36.7	0.9	31.2	0.7	0.3	15.6	27.8	3.4	14.7	6.6	0.1	0.4	0.5	0.7	1.10	3.6	3.9
		24	24.5	85.3	33.8	0.8	30.8	0.7	0.3	15.6	26.3	4.5	14.7	6.4	0.1	0.4	0.5	0.7	1.10	3.6	3.9
		25	25.3	88.5	49.1	0.8	32.3	0.7	0.3	15.7	28.6	3.7	14.7	7.6	0.1	0.5	0.4	0.9	1.10	4.0	4.4
		26	25.8	84.0	43.3	0.8	33.3	0.8	0.2	15.8	27.9	5.3	14.7	7.1	0.1	0.5	0.5	0.8	1.10	4.0	4.4
		27	26.0	86.3	46.2	0.8	33.8	0.8	0.2	15.9	29.1	4.6	14.7	7.4	0.1	0.5	0.5	0.8	1.10	4.1	4.5
		28	24.1	86.3	54.9	0.7	29.9	0.7	0.3	15.4	25.7	4.1	14.7	8.0	0.1	0.6	0.4	1.1	1.10	4.1	4.5
		29	24.4	84.8	42.4	0.8	30.6	0.7	0.3	15.5	25.9	4.7	14.7	7.1	0.1	0.5	0.5	0.9	1.10	3.8	4.2
		30	24.5	86.8	25.3	0.7	30.8	0.7	0.3	15.5	26.7	4.1	14.7	5.7	0.1	0.3	0.4	0.6	1.10	3.2	3.5

Reservoir operation for water supply to Subak irrigation schemes in Yeh Ho River Basin

Table E.1. *continued*

Month		T	RH (%)	n/N (%)	u (m/s)	ea (mbar)	w	(1 - w)	f(t)	ed (mbar)	ed - ea (mbar)	Ra (mbar)	Rs (mbar)	f (ed) (mbar)	f(n/N)	f(u)	Rn 1 (mm/day)	c	Eto* (mm/day)	Ep (mm/day)
10	October																			
	1	23.6	88.7	56.7	0.9	29.0	0.7	0.3	15.4	25.7	3.3	15.7	8.7	0.1	0.6	0.5	1.1	1.10	4.4	4.8
	2	23.7	85.3	40.2	0.6	29.1	0.7	0.3	15.4	24.8	4.3	15.7	7.3	0.1	0.5	0.4	0.9	1.10	3.9	4.2
	3	24.6	85.3	43.2	0.5	30.8	0.7	0.3	15.6	26.3	4.5	15.7	7.6	0.1	0.5	0.4	0.9	1.10	4.0	4.4
	4	22.5	90.8	37.7	0.7	26.6	0.7	0.3	15.1	24.2	2.5	15.7	7.1	0.1	0.4	0.4	0.8	1.10	3.5	3.9
	5	21.9	87.5	49.5	0.7	25.4	0.7	0.3	15.0	22.2	3.2	15.7	8.1	0.1	0.5	0.4	1.1	1.10	4.0	4.4
	6	24.2	87.0	58.8	0.9	30.1	0.7	0.3	15.5	26.2	3.9	15.7	8.9	0.1	0.6	0.5	1.1	1.10	4.6	5.0
	7	21.8	84.8	36.7	0.6	25.2	0.7	0.3	14.9	21.4	3.8	15.7	7.0	0.1	0.4	0.4	0.9	1.10	3.6	3.9
	8	24.4	89.8	56.0	0.8	30.5	0.7	0.3	15.5	27.4	3.1	15.7	8.7	0.1	0.6	0.5	1.0	1.10	4.4	4.9
	9	23.1	92.0	53.4	0.6	28.0	0.7	0.3	15.2	25.7	2.2	15.7	8.4	0.1	0.6	0.4	1.0	1.10	4.1	4.5
	10	23.8	88.3	47.2	0.7	29.2	0.7	0.3	15.4	25.8	3.4	15.7	7.9	0.1	0.5	0.4	0.9	1.10	4.1	4.5
	11	24.8	87.0	34.3	0.7	31.3	0.7	0.3	15.6	27.2	4.1	15.7	6.8	0.1	0.4	0.4	0.7	1.10	3.7	4.1
	12	24.3	82.0	32.0	0.7	30.3	0.7	0.3	15.5	24.9	5.5	15.7	6.6	0.1	0.4	0.4	0.7	1.10	3.7	4.1
	13	23.7	89.5	36.0	0.6	29.0	0.7	0.3	15.4	26.0	3.0	15.7	7.0	0.1	0.4	0.4	0.8	1.10	3.6	4.0
	14	24.3	86.5	41.2	0.6	30.3	0.7	0.3	15.5	26.2	4.1	15.7	7.4	0.1	0.5	0.4	0.8	1.10	3.9	4.3
	15	24.2	87.3	25.8	0.7	30.2	0.7	0.3	15.5	26.3	3.8	15.7	6.1	0.1	0.3	0.4	0.6	1.10	3.4	3.7
	16	24.7	85.5	50.2	0.7	31.1	0.7	0.3	15.6	26.5	4.5	15.7	8.2	0.1	0.6	0.4	1.0	1.10	4.3	4.8
	17	22.3	82.8	37.5	0.8	26.2	0.7	0.3	15.1	21.7	4.5	15.7	7.1	0.1	0.4	0.5	0.9	1.10	3.8	4.1
	18	23.6	83.8	42.9	0.7	28.8	0.7	0.3	15.3	24.2	4.7	15.7	7.6	0.1	0.5	0.4	0.9	1.10	4.0	4.4
	19	23.7	88.8	46.3	0.9	29.1	0.7	0.3	15.4	25.8	3.3	15.7	7.8	0.1	0.5	0.5	0.9	1.10	4.0	4.4
	20	24.2	88.8	32.7	0.6	30.1	0.7	0.3	15.5	26.7	3.4	15.7	6.7	0.1	0.4	0.4	0.7	1.10	3.5	3.9
	21	22.9	88.0	40.5	0.9	27.5	0.7	0.3	15.3	24.2	3.3	15.7	7.3	0.1	0.5	0.4	0.9	1.10	3.7	4.1
	22	23.4	85.5	23.6	0.7	28.5	0.7	0.3	15.3	24.3	4.1	15.7	5.9	0.1	0.3	0.5	0.6	1.10	3.3	3.7
	23	24.4	90.0	35.1	0.7	30.5	0.7	0.3	15.5	27.5	3.1	15.7	6.9	0.1	0.4	0.4	0.7	1.10	3.6	4.0
	24	23.9	94.5	35.7	0.9	29.5	0.7	0.3	15.4	27.9	1.6	15.7	6.9	0.1	0.4	0.5	0.7	1.10	3.5	3.9
	25	24.2	86.8	44.9	0.9	30.1	0.7	0.3	15.5	26.1	4.0	15.7	7.7	0.1	0.5	0.5	0.9	1.10	4.1	4.5
	26	24.6	84.3	35.1	1.5	31.0	0.7	0.3	15.6	26.1	4.9	15.7	6.9	0.1	0.4	0.6	0.7	1.10	4.1	4.5
	27	24.7	87.0	51.2	0.9	31.2	0.7	0.3	15.6	27.1	4.1	15.7	8.2	0.1	0.6	0.5	1.0	1.10	4.4	4.8
	28	25.6	86.3	38.2	1.2	32.8	0.8	0.2	15.8	28.3	4.5	15.7	7.2	0.1	0.4	0.5	0.7	1.10	4.1	4.5
	29	25.0	82.5	29.0	0.9	31.7	0.7	0.3	15.6	26.1	5.5	15.7	6.4	0.1	0.4	0.5	0.6	1.10	3.8	4.1
	30	22.8	88.5	34.6	0.9	27.3	0.7	0.3	15.2	24.1	3.1	15.7	6.8	0.1	0.4	0.5	0.8	1.10	3.6	3.9
	31	22.5	80.0	51.4	1.0	26.8	0.7	0.3	15.1	21.4	5.4	15.7	8.3	0.1	0.6	0.5	1.2	1.10	4.4	4.8

Table E.1. *continued*

Month		T	RH (%)	n/N (%)	u (m/s)	ea (mbar)	w	(1 - w)	f(t)	ed (mbar)	ed - ea (mbar)	Ra (mbar)	Rs (mbar)	f (ed) (mbar)	f(n/N)	f(u)	Rn 1 (mm/day)	c	Eto* (mm/day)	Ep (mm/day)
November	1	24.1	89.3	81.0	1.0	29.9	0.7	0.3	15.5	26.7	3.2	16.0	11.0	0.1	0.8	0.5	1.4	1.10	5.4	6.0
	2	24.1	86.0	74.3	1.0	29.9	0.7	0.3	15.5	25.7	4.2	16.0	10.4	0.1	0.8	0.5	1.4	1.10	5.3	5.8
	3	23.9	77.7	78.3	0.9	29.4	0.7	0.3	15.4	22.9	6.6	16.0	10.7	0.1	0.8	0.5	1.6	1.10	5.6	6.1
	4	23.1	83.7	63.7	0.6	27.9	0.7	0.3	15.2	23.3	4.6	16.0	9.5	0.1	0.7	0.4	1.3	1.10	4.7	5.2
	5	24.7	86.3	25.3	0.8	31.2	0.7	0.3	15.6	26.9	4.3	16.0	6.2	0.1	0.3	0.5	0.6	1.10	3.5	3.9
	6	21.8	89.3	54.3	0.6	25.3	0.7	0.3	15.0	22.6	2.7	16.0	8.7	0.1	0.6	0.4	1.2	1.10	4.1	4.6
	7	21.2	77.3	44.0	0.7	24.1	0.7	0.3	14.8	18.6	5.5	16.0	7.8	0.2	0.5	0.4	1.1	1.10	4.0	4.4
	8	20.6	86.7	66.5	0.7	22.8	0.7	0.3	14.7	19.8	3.0	16.0	9.7	0.1	0.7	0.4	1.5	1.10	4.5	4.9
	9	22.6	76.3	50.3	0.6	26.8	0.7	0.3	15.1	20.5	6.3	16.0	8.3	0.1	0.6	0.4	1.2	1.10	4.4	4.8
	10	20.8	85.0	46.7	0.5	23.3	0.7	0.3	14.7	19.8	3.5	16.0	8.0	0.1	0.5	0.4	1.1	1.10	3.9	4.2
	11	22.0	88.7	23.0	0.8	25.7	0.7	0.3	15.0	22.8	2.9	16.0	6.0	0.1	0.3	0.4	0.6	1.10	3.1	3.5
	12	21.6	93.3	43.3	0.6	24.8	0.7	0.3	14.9	23.2	1.7	16.0	7.7	0.1	0.5	0.4	0.9	1.10	3.7	4.0
	13	22.7	85.7	30.5	0.9	27.2	0.7	0.3	15.2	23.3	3.9	16.0	6.6	0.1	0.4	0.5	0.7	1.10	3.6	3.9
	14	21.0	86.7	36.0	0.6	23.6	0.7	0.3	14.8	20.4	3.1	16.0	7.1	0.1	0.4	0.4	0.9	1.10	3.5	3.9
	15	22.5	85.3	49.0	0.6	26.7	0.7	0.3	15.1	22.8	3.9	16.0	8.2	0.1	0.5	0.4	1.1	1.10	4.1	4.5
	16	21.6	91.3	25.0	0.6	24.8	0.7	0.3	14.9	22.6	2.1	16.0	6.1	0.1	0.3	0.4	0.6	1.10	3.1	3.4
	17	23.1	93.3	21.7	0.4	27.9	0.7	0.3	15.2	26.0	1.9	16.0	5.9	0.1	0.3	0.4	0.5	1.10	3.0	3.3
	18	22.6	88.3	36.0	0.5	26.9	0.7	0.3	15.1	23.8	3.1	16.0	7.1	0.1	0.4	0.4	0.8	1.10	3.6	4.0
	19	23.0	91.0	13.7	0.7	27.6	0.7	0.3	15.2	25.1	2.5	16.0	5.2	0.1	0.2	0.4	0.4	1.10	2.8	3.1
	20	21.7	91.3	29.5	0.5	25.0	0.7	0.3	14.9	22.9	2.2	16.0	6.5	0.1	0.4	0.4	0.7	1.10	3.2	3.6
	21	22.9	92.7	18.5	0.4	26.5	0.7	0.3	15.2	25.5	2.0	16.0	5.6	0.1	0.3	0.4	0.5	1.10	2.9	3.2
	22	22.4	83.3	20.5	0.9	26.5	0.7	0.3	15.1	22.1	4.4	16.0	5.8	0.1	0.3	0.5	0.6	1.10	3.3	3.6
	23	21.3	85.0	47.7	0.6	24.3	0.7	0.3	14.9	20.7	3.6	16.0	8.1	0.1	0.5	0.4	1.1	1.10	4.0	4.4
	24	21.7	86.3	57.0	0.6	25.1	0.7	0.3	14.9	21.7	3.4	16.0	8.9	0.1	0.6	0.4	1.2	1.10	4.3	4.7
	25	21.9	82.3	47.7	0.5	25.4	0.7	0.3	15.0	20.9	4.5	16.0	8.1	0.1	0.5	0.4	1.1	1.10	4.0	4.4
	26	23.0	81.3	51.5	0.6	27.7	0.7	0.3	15.2	22.5	5.2	16.0	8.4	0.1	0.6	0.4	1.1	1.10	4.3	4.8
	27	20.5	85.3	48.0	0.4	22.7	0.7	0.3	14.7	19.4	3.3	16.0	8.1	0.1	0.5	0.4	1.1	1.10	3.8	4.2
	28	21.3	90.3	22.7	0.6	24.2	0.7	0.3	14.8	21.8	2.3	16.0	5.9	0.1	0.3	0.4	0.6	1.10	3.0	3.3
	29	22.5	84.0	35.7	0.7	26.6	0.7	0.3	15.1	22.3	4.3	16.0	7.1	0.1	0.4	0.4	0.8	1.10	3.7	4.1
	30	22.9	84.0	36.7	0.7	27.5	0.7	0.3	15.2	23.1	4.4	16.0	7.1	0.1	0.4	0.4	0.8	1.10	3.8	4.2

Reservoir operation for water supply to Subak irrigation schemes in Yeh Ho River Basin

Table E.1. *continued*

Month		T	RH (%)	n/N (%)	u (m/s)	ea (mbar)	w	(1 - w)	f(t)	ed (mbar)	ed - ea (mbar)	Ra (mbar)	Rs (mbar)	f (ed) (mbar)	f(n/N)	f(u)	Rn 1 (mm/day)	c	Eto* (mm/day)	Ep (mm/day)
December	1	22.8	86.5	23.9	0.5	27.2	0.7	0.3	15.2	23.5	3.7	16.0	6.1	0.1	0.3	0.4	0.6	1.10	3.2	3.6
	2	22.1	88.0	25.2	0.8	25.8	0.7	0.3	15.0	22.7	3.1	16.0	6.2	0.1	0.3	0.5	0.6	1.10	3.3	3.6
	3	22.3	87.0	23.3	0.7	26.2	0.7	0.3	15.1	22.8	3.4	16.0	6.0	0.1	0.3	0.4	0.6	1.10	3.2	3.5
	4	22.3	88.7	36.0	0.7	26.2	0.7	0.3	15.1	23.2	3.0	16.0	7.1	0.1	0.4	0.4	0.8	1.10	3.6	4.0
	5	22.8	91.3	33.9	0.5	27.2	0.7	0.3	15.2	24.8	2.4	16.0	6.9	0.1	0.4	0.4	0.7	1.10	3.5	3.8
	6	22.1	84.0	42.3	0.5	26.0	0.7	0.3	15.0	21.8	4.2	16.0	7.7	0.1	0.5	0.4	1.0	1.10	3.9	4.3
	7	20.6	85.7	36.4	0.4	22.9	0.7	0.3	14.7	19.6	3.3	16.0	7.1	0.1	0.4	0.4	0.9	1.10	3.5	3.8
	8	20.6	83.7	21.2	0.6	22.8	0.7	0.3	14.7	19.1	3.7	16.0	5.8	0.1	0.3	0.4	0.6	1.10	3.1	3.4
	9	23.4	80.0	33.1	0.7	28.6	0.7	0.3	15.3	22.9	5.7	16.0	6.9	0.1	0.4	0.4	0.8	1.10	3.8	4.2
	10	22.6	85.7	18.1	1.1	26.9	0.7	0.3	15.1	23.0	3.9	16.0	5.6	0.1	0.3	0.5	0.5	1.10	3.2	3.5
	11	22.5	83.7	30.3	0.6	26.8	0.7	0.3	15.1	22.4	4.4	16.0	6.6	0.1	0.4	0.4	0.7	1.10	3.5	3.9
	12	22.2	87.7	30.3	1.2	26.1	0.7	0.3	15.0	22.9	3.2	16.0	6.6	0.1	0.4	0.5	0.7	1.10	3.4	3.8
	13	23.0	77.0	32.8	1.2	27.7	0.7	0.3	15.2	21.3	6.4	16.0	6.8	0.1	0.4	0.5	0.8	1.10	4.1	4.5
	14	22.4	89.3	33.4	0.5	26.5	0.7	0.3	15.1	23.7	2.8	16.0	6.9	0.1	0.4	0.4	0.8	1.10	3.5	3.8
	15	21.5	91.0	18.5	0.7	24.6	0.7	0.3	14.9	22.4	2.2	16.0	5.6	0.1	0.3	0.4	0.5	1.10	2.9	3.2
	16	22.1	84.3	8.5	0.7	25.9	0.7	0.3	15.0	21.9	4.1	16.0	4.7	0.1	0.2	0.4	0.4	1.10	2.8	3.1
	17	22.6	91.0	17.6	1.1	26.9	0.7	0.3	15.1	24.5	2.4	16.0	5.5	0.1	0.3	0.5	0.5	1.10	3.0	3.3
	18	23.0	84.7	24.9	0.7	27.6	0.7	0.3	15.2	23.4	4.2	16.0	6.1	0.1	0.3	0.4	0.6	1.10	3.4	3.7
	19	23.1	84.3	17.4	1.0	28.0	0.7	0.3	15.2	23.6	4.4	16.0	5.5	0.1	0.3	0.5	0.5	1.10	3.2	3.6
	20	22.4	84.3	21.7	0.7	26.5	0.7	0.3	15.1	22.3	4.2	16.0	5.9	0.1	0.4	0.4	0.6	1.10	3.3	3.6
	21	22.0	85.3	29.8	0.9	25.6	0.7	0.3	15.0	21.9	3.8	16.0	6.6	0.1	0.4	0.5	0.7	1.10	3.5	3.8
	22	23.2	84.0	29.1	0.6	28.1	0.7	0.3	15.3	23.6	4.5	16.0	6.5	0.1	0.4	0.4	0.7	1.10	3.6	3.9
	23	22.2	84.3	28.2	0.6	26.2	0.7	0.3	15.1	22.1	4.1	16.0	6.4	0.1	0.4	0.4	0.7	1.10	3.4	3.8
	24	22.1	85.7	36.6	0.8	25.9	0.7	0.3	15.0	22.2	3.7	16.0	7.2	0.1	0.4	0.5	0.9	1.10	3.7	4.1
	25	21.9	89.3	21.5	0.7	25.5	0.7	0.3	15.0	22.7	2.7	16.0	5.9	0.1	0.3	0.4	0.6	1.10	3.1	3.4
	26	21.9	90.0	40.8	1.0	25.5	0.7	0.3	15.0	22.9	2.5	16.0	7.5	0.1	0.5	0.5	0.9	1.10	3.7	4.1
	27	22.3	80.7	30.2	0.6	26.2	0.7	0.3	15.1	21.1	5.1	16.0	6.6	0.1	0.4	0.4	0.8	1.10	3.6	4.0
	28	22.9	80.0	32.4	1.0	27.5	0.7	0.3	15.2	22.0	5.5	16.0	6.8	0.1	0.4	0.5	0.8	1.10	3.9	4.3
	29	22.8	88.0	35.7	0.6	27.2	0.7	0.3	15.2	24.0	3.3	16.0	7.1	0.1	0.4	0.4	0.8	1.10	3.6	4.0
	30	23.1	91.7	34.9	0.5	27.9	0.7	0.3	15.2	25.6	2.3	16.0	7.0	0.1	0.4	0.4	0.7	1.10	3.5	3.9
	31	23.2	83.0	35.8	1.0	28.0	0.7	0.3	15.3	23.3	4.8	16.0	7.1	0.1	0.4	0.5	0.8	1.10	3.9	4.3

Annex F. Hydraulic profile of outlets

Table F.1. Hydraulic profile of the outlets of Telaga Tunjung Reservoir for Meliling Subak irrigation scheme and for domestic water supply

El. NWL	El. Dead	Open	A1	Q1 (Meliling)	%	V1	A2	Q2 (Water supply)	%	V2	A3	Q3 (Jetflow)	%	V3	Q Total	Qmax
m	m	m		m³/s		m/s		m³/s		m/s		m³/s		m/s		
190.70	190	0.70		0.94		4.78		0.15		4.78	0.38	1.84	100	4.78	2.93	1.84
190.65	190	0.65		0.91		4.61		0.14		4.61	0.37	1.72	93	4.46	2.77	1.84
190.60	190	0.60		0.87		4.43		0.14		4.43	0.35	1.56	86	4.04	2.56	1.84
190.55	190	0.55		0.83		4.24		0.13		4.24	0.32	1.38	79	3.57	2.34	1.84
190.50	190	0.50	0.20	0.79	100	4.04		0.13		4.04	0.29	1.19	71	3.09	2.11	1.84
190.45	190	0.45	0.19	0.71	90	3.63		0.12		3.84	0.26	1.00	64	2.61	1.84	1.84
190.40	190	0.40	0.17	0.61	80	3.10		0.11		3.62	0.23	0.82	57	2.14	1.54	1.84
190.35	190	0.35	0.15	0.50	70	2.53		0.11		3.38	0.19	0.65	50	1.69	1.25	1.84
190.30	190	0.30	0.12	0.39	60	1.96		0.10		3.13	0.16	0.49	43	1.28	0.98	1.84
190.25	190	0.25	0.10	0.28	50	1.43		0.09		2.86	0.12	0.35	36	0.92	0.72	1.84
190.20	190	0.20	0.07	0.19	40	0.95	0.03	0.08	100	2.56	0.09	0.23	29	0.60	0.50	1.84
190.15	190	0.15	0.05	0.11	30	0.56	0.03	0.06	75	1.78	0.06	0.13	21	0.35	0.30	1.84
190.10	190	0.10	0.03	0.05	20	0.26	0.02	0.03	50	0.90	0.03	0.06	14	0.16	0.14	1.84
190.05	190	0.05	0.01	0.01	10	0.07	0.01	0.01	25	0.25	0.01	0.02	7	0.04	0.04	1.84
190.00	190	0.00	0.00	0.00	0	0.00	0.00	0.00	0	0.00	0.00	0.00	0	0.00	0.00	1.84

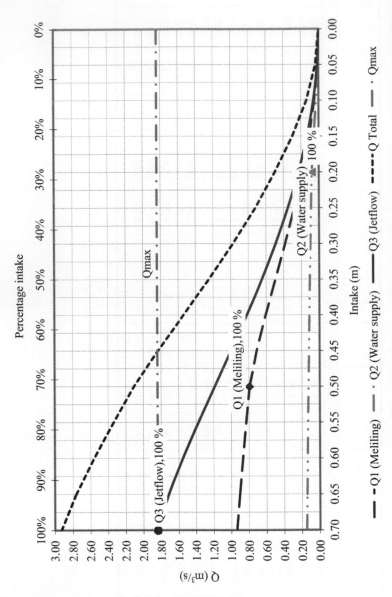

Figure F.2. Water release from Telaga Tunjung Reservoir in relation to the percentage of opening of the outlets

Annex G. Reservoir and its hydraulic structures

Reservoir:

River basin	$= 81.5 \text{ km}^2$
Total storage	$= 1{,}260{,}000 \text{ m}^3$
Effective storage	$= 1{,}000{,}000 \text{ m}^3$
Death storage (50 yr)	$= 261{,}000 \text{ m}^3$
Reservoir storage area	$= 16.5 \text{ ha}$
Flood discharge design (Q 1.5 PMF)	$= 776 \text{ m}^3/\text{s}$
Maximum water level (Q 1.5 PMF)	$= 201.51 \text{ m +MSL}$
Normal water level	$= 199.00 \text{ m +MSL}$
Minimum water level	$= 190.70 \text{ m +MSL}$

Main dam:

Type	= Earthfill dam (zonal random with vertical core)
Maximum height	= 33 m
Length of top dam	= 225.4 m
Volume of bank (including cofferdam)	$= 246{,}632 \text{ m}^3$
Free board (Q20 yr)	= 1.49 m

Diversion:

Type	= Conduit
Length	$= 2 @ 3.5 \times 3.5 \text{ m } [193 \text{ m}]$
Top cofferdam elevation	= 188.00 m +MSL
Free board (Q20 yr)	= 0.43 m
Flood discharge design (Q20 yr)	$= 357 \text{ m}^3/\text{s}$

Overflow:

Location	= abutment in the right
Type	= side canal weir without gate and steep canal
Stilling basin	= USBR Type I with end-sill
Top weir elevation	= 199.00 m +MSL
Maximum Water Level	= 201.51 m +MSL
Width and length overflow	= 13 m – 27 m and 93 m
Width and length steep canal	= 27 m and 62.85 m

Intake:

Type	= Morning glory with steel sliding gate and bonnet type steel pipes with valves (steel conduit + valve)
Capacity	= 1.87 m³/s [0.141 + 1.725]

Annex H. Information on reservoir sedimentation

Borland and Miller (1958): Area Reduction Method

Table H.1. Standard types of reservoirs

Classification	Type of reservoir	M	C	m	n
I	*Lake*	3.5 - 4.5	3.42	1.5	0.2
II	*Floodplain-foothill*	2.5 - 3.5	2.32	0.5	0.4
III	*Hill*	1.5 - 2.5	15.9	1.1	2.3
IV	*Gorge*	1.0 - 1.5	4.23	0.1	2.5

M = *slope of curve*

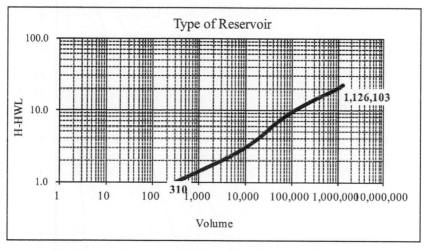

Figure H.1. Gradual sediment increase in a floodplain-foothill type of reservoir

The type of reservoir based on Table H.1 is:

X	3.56	Log of highest minus lowest capacity (X on graph)
Y	1.34	Log of highest minus lowest H level (Y on graph)
M (X/Y)	2.65	Type II - slope

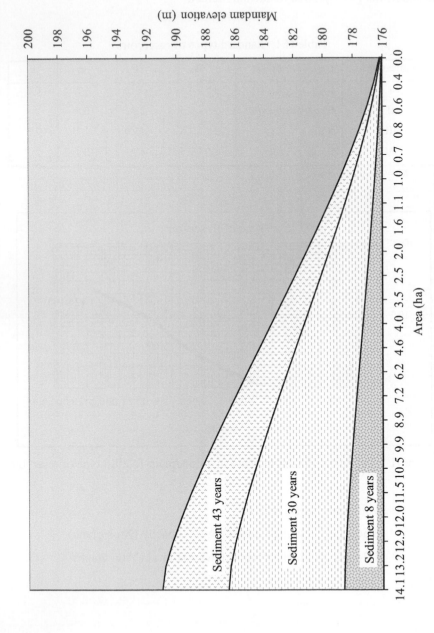

Figure H.2. Sedimentation profile of Telaga Tunjung Reservoir

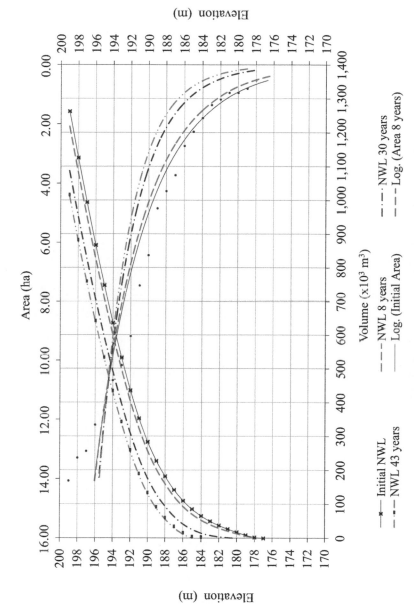

Figure H.3. Reservoir capacity for the Telaga Tunjung Reservoir based on sedimentation

Annex I. Infiltration and percolation

Table I.1. Water requirements for the pre-saturation of the field and water layer S
('Van der Goor' method)

Eo+P	T = 30 days		T = 45 days	
mm/day	S = 250 mm	S = 300 mm	S = 250 mm	S = 300 mm
5.0	11.1	12.7	8.4	9.5
5.5	11.4	13.0	8.8	9.8
6.0	11.7	13.3	9.1	10.1
6.5	12.0	13.6	9.4	10.4
7.0	12.3	13.9	9.8	10.8
7.5	12.6	14.2	10.1	11.1
8.0	13.0	14.5	10.5	11.4
8.5	13.3	14.8	10.8	11.8
9.0	13.6	15.2	11.2	12.1
9.5	14.0	15.5	11.6	12.5
10.0	14.3	15.8	12.0	12.9
10.5	14.7	16.2	12.4	13.2
11.0	15.0	16.5	12.8	13.6

Source: Irrigation Planning Criteria (KP) 01

Annex J. Results of the measurements in a paddy terraces block

Table J.1. Observations at inlet and outlet of a paddy terraces block

2013	WDU (Tektek)		Drain (Pengutangan)		Δ	%
	m³/s		m³/s			
April	0.026		0.019			
	0.024		0.019			
	0.024		0.029			
	0.023		0.016			
	0.024		0.019			
	0.024		0.022			
	0.027		0.019			
	0.025		0.026			
	0.023		0.029			
	0.024		0.034			
	0.026		0.034			
	0.026		0.035			
	0.025		0.031			
	0.028		0.031			
	0.033		0.032			
	0.033		0.032			
	0.033		0.032			
	0.033		0.032			
	0.033		0.032			
	0.086		0.086			
	0.086		0.086			
	0.086		0.086			
	0.081		0.039			
	0.069		0.033			
	0.069		0.033			
	0.069		0.033			
	0.069		0.033			
	0.069		0.033			
	0.065		0.021			
	0.065	0.044	0.021	0.034	-0.010	-29%
Mei	0.065		0.021			
	0.065		0.021			
	0.051		0.036			
	0.044		0.041			
	0.044		0.041			

Table J.1. *continued*

2013	WDU (Tektek) m³/s	Drain (Pengutangan) m³/s	Δ	%
	0.038	0.058		
	0.031	0.063		
	0.031	0.063		
	0.031	0.063		
	0.029	0.068		
	0.029	0.068		
	0.029	0.068		
	0.030	0.071		
	0.032	0.059		
	0.032	0.075		
	0.035	0.058		
	0.032	0.077		
	0.032	0.077		
	0.032	0.077		
	0.030	0.074		
	0.032	0.080		
	0.032	0.082		
	0.033	0.090		
	0.000	0.090		
	0.000	0.090		
	0.000	0.090		
	0.031	0.086		
	0.032	0.086		
	0.034	0.107		
	0.034	0.084		
	0.034	0.032	0.084	0.0694	0.0370	53%
June	0.034	0.084		
	0.034	0.084		
	0.039	0.084		
	0.039	0.084		
	0.040	0.092		
	0.040	0.092		
	0.040	0.092		
	0.041	0.092		
	0.041	0.092		
	0.043	0.107		
	0.043	0.107		

Table J.1. *continued*

2013	WDU (*Tektek*) m³/s		Drain (*Pengutangan*) m³/s		Δ	%
	0.043		0.107			
	0.043		0.107			
	0.044		0.115			
	0.044		0.115			
	0.044		0.115			
	0.044		0.115			
	0.044		0.137			
	0.044		0.137			
	0.044		0.137			
	0.044		0.137			
	0.008		0.137			
	0.000		0.137			
	0.000		0.140			
	0.027		0.140			
	0.027		0.140			
	0.027		0.140			
	0.044		0.140			
	0.044		0.140			
	0.042	0.036	0.140	0.116	0.080	69%
July	0.042		0.140			
	0.042		0.140			
	0.032		0.103			
	0.032		0.103			
	0.032		0.103			
	0.004		0.103			
	0.004		0.103			
	0.004		0.103			
	0.028		0.086			
	0.028		0.086			
	0.028		0.086			
	0.028		0.086			
	0.028		0.086			
	0.028		0.086			
	0.003		0.079			
	0.003		0.079			

Table J.1. *continued*

2013	WDU (*Tektek*)		Drain (*Pengutangan*)	Δ	%
	m³/s		m³/s		
August	0.030		0.079		
	0.030		0.079		
	0.030		0.079		
	0.030		0.079		
	0.025		0.077		
	0.025		0.077		
	0.025		0.077		
	0.026		0.077		
	0.029		0.077		
	0.029		0.077		
	0.029		0.077		
	0.029		0.077		
	0.023	0.025	0.078	0.088	0.064	72%
	0.023		0.078		
	0.023		0.078		
	0.023		0.078		
	0.023		0.078		
	0.021		0.077		
	0.021		0.077		
	0.021		0.077		
	0.028		0.077		
	0.028		0.077		
	0.028		0.077		
	0.028		0.077		
	0.021		0.074		
	0.021		0.074		
	0.021		0.074		
	0.026		0.074		
	0.026		0.074		
	0.026		0.074		
	0.026		0.074		
	0.021		0.074		
	0.021		0.074		
	0.021		0.074		
	0.021		0.074		
	0.017		0.075		
	0.017		0.075		

Table J.1. *continued*

2013	WDU (*Tektek*)		Drain (*Pengutangan*)		Δ	%
	m³/s		m³/s			
	0.017		0.075			
	0.017		0.075			
	0.017		0.075			
	0.017		0.075			
	0.018		0.073			
	0.018		0.073			
	0.018	0.022	0.073	0.075	0.054	71%
September	0.018		0.073			
	0.024		0.073			
	0.026		0.034			
	0.027		0.034			
	0.027		0.023			
	0.028		0.023			
	0.028		0.023			
	0.029		0.023			
	0.028		0.033			
	0.027		0.032			
	0.027		0.032			
	0.029		0.019			
	0.031		0.016			
	0.031		0.022			
	0.028		0.022			
	0.029		0.022			
	0.029		0.031			
	0.033		0.031			
	0.031		0.031			
	0.032		0.034			
	0.030		0.034			
	0.030		0.032			
	0.030		0.025			
	0.029		0.025			
	0.030		0.025			
	0.030		0.033			
	0.025		0.033			
	0.025		0.033			
	0.025		0.020			
	0.025	0.028	0.020	0.030	0.002	8%

Table J.1. *continued*

2013	WDU (*Tektek*)		Drain (*Pengutangan*)	Δ	%
	m³/s		m³/s		
October	0.025		0.020		
	0.024		0.020		
	0.029		0.024		
	0.029		0.024		
	0.029		0.024		
	0.028		0.024		
	0.028		0.024		
	0.028		0.024		
	0.028		0.024		
	0.026		0.032		
	0.026		0.032		
	0.028		0.023		
	0.029		0.019		
	0.030		0.020		
	0.030		0.020		
	0.030		0.034		
	0.029		0.034		
	0.028		0.034		
	0.028		0.034		
	0.028		0.034		
	0.030		0.029		
	0.030		0.029		
	0.030		0.029		
	0.030		0.029		
	0.012		0.019		
	0.012		0.019		
	0.012		0.019		
	0.026		0.026		
	0.026		0.026		
	0.027		0.026		
	0.028	0.027	0.030	0.026	-0.001	-2%
November	0.028		0.030		
	0.028		0.030		
	0.029		0.031		
	0.029		0.031		
	0.029		0.032		
	0.029		0.032		

Table J.1. *continued*

2013	WDU (*Tektek*)		Drain (*Pengutangan*)		Δ	%
	m³/s		m³/s			
	0.029		0.033			
	0.029		0.026			
	0.027		0.026			
	0.028		0.029			
	0.028		0.032			
	0.030		0.032			
	0.030		0.032			
	0.030		0.019			
	0.027		0.024			
	0.027		0.024			
	0.026		0.024			
	0.025		0.023			
	0.025		0.023			
	0.025		0.023			
	0.024		0.070			
	0.026		0.070			
	0.026		0.070			
	0.024		0.058			
	0.023		0.077			
	0.021		0.077			
	0.022		0.061			
	0.022		0.075			
	0.022		0.075			
	0.022	0.026	0.075	0.042	0.016	37%
December	0.022		0.075			
	0.022		0.075			
	0.024		0.076			
	0.024		0.071			
	0.024		0.071			
	0.025		0.069			
	0.025		0.058			
	0.025		0.058			
	0.028		0.058			
	0.028		0.057			
	0.028		0.057			
	0.029		0.074			
	0.030		0.072			

Table J.1. *continued*

2014	WDU (*Tektek*)		Drain (*Pengutangan*)		Δ	%
	m³/s		m³/s			
	0.030		0.072			
	0.028		0.074			
	0.025		0.074			
	0.026		0.074			
	0.025		0.077			
	0.025		0.070			
	0.025		0.070			
	0.024		0.075			
	0.025		0.075			
	0.025		0.077			
	0.023		0.077			
	0.023		0.077			
	0.025		0.059			
	0.021		0.057			
	0.021		0.057			
	0.021		0.057			
	0.021		0.057			
	0.021	0.025	0.057	0.068	0.043	63%
January	0.021		0.057			
	0.021		0.057			
	0.000		0.057			
	0.000		0.057			
	0.000		0.057			
	0.026		0.039			
	0.026		0.039			
	0.026		0.039			
	0.035		0.046			
	0.035		0.046			
	0.035		0.046			
	0.039		0.046			
	0.039		0.069			
	0.039		0.069			
	0.039		0.072			
	0.039		0.072			
	0.039		0.072			
	0.039		0.072			
	0.039		0.078			

Table J.1. *continued*

2014	WDU (*Tektek*)		Drain (*Pengutangan*)		Δ	%
	m³/s		m³/s			
	0.039		0.078			
	0.039		0.078			
	0.040		0.071			
	0.040		0.071			
	0.040		0.071			
	0.040		0.075			
	0.040		0.075			
	0.048		0.083			
	0.048		0.083			
	0.048		0.083			
	0.027		0.075			
	0.027	0.033	0.075	0.065	0.032	50%
February	0.027		0.075			
	0.027		0.075			
	0.033		0.085			
	0.033		0.085			
	0.033		0.085			
	0.036		0.087			
	0.036		0.087			
	0.036		0.087			
	0.036		0.087			
	0.032		0.085			
	0.032		0.085			
	0.032		0.085			
	0.027		0.078			
	0.027		0.078			
	0.027		0.078			
	0.020		0.078			
	0.020		0.078			
	0.000		0.078			
	0.000		0.078			
	0.000		0.078			
	0.023		0.076			
	0.023		0.076			
	0.023		0.076			
	0.023		0.076			
	0.025		0.065			

Table J.1. *continued*

2014	WDU (*Tektek*)		Drain (*Pengutangan*)		Δ	%
	m³/s		m³/s			
March	0.025		0.065			
	0.025		0.065			
	0.025	0.025	0.065	0.078	0.053	68%
	0.025		0.065			
	0.025		0.065			
	0.020		0.064			
	0.020		0.064			
	0.020		0.064			
	0.016		0.069			
	0.016		0.069			
	0.016		0.069			
	0.016		0.069			
	0.019		0.076			
	0.019		0.076			
	0.019		0.076			
	0.019		0.076			
	0.019		0.076			
	0.019		0.076			
	0.019		0.076			
	0.013		0.078			
	0.013		0.078			
	0.013		0.078			
	0.012		0.075			
	0.012		0.075			
	0.012		0.075			
	0.012		0.075			
	0.019		0.078			
	0.019		0.078			
	0.019		0.078			
	0.019		0.078			
	0.018		0.084			
	0.018		0.084			
	0.018		0.084			
	0.018	0.017	0.084	0.075	0.057	77%
April	0.018		0.084			

Table J.1. *continued*

2014	WDU (*Tektek*)		Drain (*Pengutangan*)		Δ	%
	m³/s		m³/s			
April	0.018		0.084			
	0.000		0.089			
	0.000		0.089			
	0.000		0.089			
	0.000		0.093			
	0.000		0.093			
	0.000		0.093			
	0.000		0.107			
	0.000		0.107			
	0.000		0.107			
	0.000		0.107			
	0.000		0.094			
	0.006		0.094			
	0.011		0.080			
	0.011		0.080			
	0.006		0.060			
	0.006		0.060			
	0.006		0.052			
	0.006		0.052			
	0.005		0.038			
	0.005		0.038			
	0.004		0.033			
	0.004		0.033			
	0.004		0.025			
	0.004		0.025			
	0.004		0.016			
	0.004		0.016			
	0.004		0.012			
	0.004		0.012			
	0.002	0.004	0.009	0.063	0.059	94%
Mei	0.002		0.009			
	0.001		0.002			
	0.001		0.002			
	0.001		0.000			
	0.000		0.000			

Table J.1. *continued*

2014	WDU (*Tektek*)		Drain (*Pengutangan*)	Δ	%
	m³/s		m³/s		
	0.000		0.000		
	0.000		0.000		
	0.000		0.000		
	0.000		0.000		
	0.000		0.000		
	0.000		0.000		
	0.000		0.000		
	0.000		0.000		
	0.000		0.000		
	0.000		0.000		
	0.000		0.000		
	0.000		0.000		
	0.000		0.000		
	0.000		0.000		
	0.000		0.000		
	0.000		0.000		
	0.000		0.000		
	0.000		0.000		
	0.000		0.000		
	0.000		0.000		
	0.000		0.000		
	0.000		0.000		
	0.000		0.000		
	0.000		0.000		
	0.000		0.000		
	0.000	0.00021	0.000	0.0004	0.0002	49%
June	0.000		0.000		
	0.000		0.000		
	0.000		0.000		
	0.000		0.000		
	0.000		0.000		
	0.000		0.000		
	0.000		0.000		
	0.000		0.000		
	0.000		0.000		
	0.000		0.000		
	0.000		0.000		

Table J.1. *continued*

2014	WDU (*Tektek*)		Drain (*Pengutangan*)		Δ	%
	m³/s		m³/s			
	0.000		0.000			
	0.000		0.000			
	0.000		0.000			
	0.000		0.000			
	0.000		0.000			
	0.000		0.000			
	0.000		0.000			
	0.000		0.000			
	0.000		0.000			
	0.000		0.000			
	0.000		0.000			
	0.000		0.000			
	0.000		0.000			
	0.000		0.000			
	0.000		0.000			
	0.000		0.000			
	0.000		0.000			
	0.000		0.000			
	0.000	0.000	0.000	0.000	0.000	0%
July	0.000		0.000			
	0.000		0.000			
	0.000		0.000			
	0.000		0.000			
	0.000		0.000			
	0.000		0.000			
	0.000		0.000			
	0.000		0.000			
	0.000		0.000			
	0.000		0.000			
	0.000		0.000			
	0.000		0.000			
	0.000		0.000			
	0.000		0.000			
	0.000		0.000			
	0.000		0.000			
	0.000		0.000			
	0.000		0.000			

Table J.1. *continued*

2014	WDU (*Tektek*)		Drain (*Pengutangan*)		Δ	%
	m³/s		m³/s			
	0.000		0.000			
	0.000		0.000			
	0.000		0.000			
	0.000		0.000			
	0.000		0.000			
	0.000		0.000			
	0.000		0.000			
	0.000		0.000			
	0.000		0.000			
	0.000		0.000			
	0.000		0.000			
	0.000		0.000			
	0.000	0.000	0.000	0.000	0.000	0%
August	0.000		0.000			
	0.000		0.000			
	0.000		0.000			
	0.000		0.000			
	0.000		0.000			
	0.000		0.000			
	0.000		0.000			
	0.000		0.000			
	0.000		0.000			
	0.000		0.000			
	0.000		0.000			
	0.000		0.000			
	0.000		0.000			
	0.000		0.000			
	0.000		0.000			
	0.000		0.000			
	0.000		0.000			
	0.000		0.000			
	0.000		0.000			
	0.000		0.000			
	0.000		0.000			
	0.000		0.000			
	0.000		0.000			
	0.000		0.000			

Table J.1. *continued*

2014	WDU (*Tektek*)		Drain (*Pengutangan*)		Δ	%
	m³/s		m³/s			
September	0.000		0.000			
	0.000		0.000			
	0.000		0.000			
	0.000		0.000			
	0.000		0.000			
	0.000		0.000			
	0.000	0.000	0.000	0.000	0.000	0%
	0.000		0.000			
	0.000		0.000			
	0.000		0.000			
	0.000		0.000			
	0.000		0.000			
	0.000		0.000			
	0.000		0.000			
	0.000		0.000			
	0.000		0.000			
	0.000		0.000			
	0.000		0.000			
	0.000		0.000			
	0.000		0.000			
	0.000		0.000			
	0.000		0.000			
	0.000		0.000			
	0.000		0.000			
	0.000		0.000			
	0.000		0.000			
	0.000		0.000			
	0.000		0.000			
	0.000		0.000			
	0.000		0.000			
	0.000		0.000			
	0.000		0.000			
	0.000		0.000			
	0.000		0.000			
	0.000		0.000			
	0.000		0.000			
	0.000		0.000			
	0.000		0.000			
	0.000	0.000	0.000	0.000	0.000	0%

Table J.1. *continued*

2014	WDU (*Tektek*)		Drain (*Pengutangan*)	Δ	%
	m³/s		m³/s		
October	0.000		0.000		
	0.000		0.000		
	0.000		0.000		
	0.000		0.000		
	0.000		0.000		
	0.000		0.000		
	0.000		0.000		
	0.000		0.000		
	0.000		0.000		
	0.000		0.000		
	0.000		0.000		
	0.000		0.000		
	0.000		0.000		
	0.000		0.000		
	0.000		0.000		
	0.000		0.000		
	0.000		0.000		
	0.000		0.000		
	0.000		0.000		
	0.000		0.000		
	0.000		0.000		
	0.000		0.000		
	0.000		0.000		
	0.000		0.000		
	0.000		0.000		
	0.000		0.000		
	0.000		0.000		
	0.000		0.000		
	0.000		0.000		
	0.000		0.000		
	0.000		0.000		
	0.000	0.000	0.000	0.000	0.000	0%
November	0.000		0.000		
	0.000		0.000		
	0.000		0.000		
	0.000		0.000		
	0.000		0.000		
	0.000		0.000		

Table J.1. *continued*

2014	WDU (*Tektek*)		Drain (*Pengutangan*)		Δ	%
	m^3/s		m^3/s			
	0.000		0.000			
	0.000		0.000			
	0.000		0.000			
	0.000		0.000			
	0.000		0.000			
	0.000		0.000			
	0.000		0.000			
	0.000		0.000			
	0.000		0.000			
	0.000		0.000			
	0.070		0.000			
	0.070		0.000			
	0.077		0.000			
	0.077		0.000			
	0.081		0.000			
	0.081		0.000			
	0.083		0.000			
	0.083		0.000			
	0.087		0.000			
	0.087		0.000			
	0.087		0.000			
	0.087		0.000			
	0.089		0.000			
	0.089	0.038	0.000	0.000	-0.038	0%
December	0.087		0.106			
	0.087		0.106			
	0.098		0.129			
	0.098		0.129			
	0.098		0.129			
	0.098		0.129			
	0.103		0.127			
	0.103		0.127			
	0.096		0.115			
	0.096		0.115			
	0.095		0.171			
	0.095		0.171			
	0.095		0.171			

Table J.1. *continued*

2015	WDU (*Tektek*)		Drain (*Pengutangan*)		Δ	%
	m³/s		m³/s			
	0.095		0.171			
	0.095		0.171			
	0.095		0.171			
	0.095		0.171			
	0.095		0.171			
	0.088		0.123			
	0.088		0.123			
	0.088		0.123			
	0.088		0.123			
	0.095		0.131			
	0.095		0.131			
	0.095		0.131			
	0.095		0.131			
	0.095		0.131			
	0.095		0.131			
	0.095		0.183			
	0.095		0.183			
	0.095	0.095	0.183	0.142	0.047	33%
January	0.095		0.183			
	0.095		0.183			
	0.095		0.184			
	0.095		0.184			
	0.103		0.188			
	0.103		0.188			
	0.096		0.168			
	0.096		0.168			
	0.095		0.165			
	0.095		0.165			
	0.095		0.165			
	0.043		0.158			
	0.043		0.158			
	0.043		0.158			
	0.043		0.158			
	0.044		0.183			
	0.044		0.183			
	0.044		0.183			
	0.044		0.183			

Table J.1. *continued*

2015	WDU (*Tektek*)		Drain (*Pengutangan*)		Δ	%
	m³/s		m³/s			
	0.044		0.200			
	0.044		0.200			
	0.038		0.200			
	0.044		0.200			
	0.008		0.212			
	0.008		0.212			
	0.008		0.212			
	0.027		0.187			
	0.027		0.187			
	0.027		0.187			
	0.044		0.187			
	0.044	0.057	0.187	0.183	0.126	69%
February	0.044		0.187			
	0.044		0.187			
	0.044		0.187			
	0.032		0.162			
	0.032		0.162			
	0.032		0.162			
	0.017		0.157			
	0.017		0.157			
	0.017		0.157			
	0.032		0.162			
	0.032		0.162			
	0.032		0.162			
	0.027		0.196			
	0.027		0.196			
	0.027		0.196			
	0.020		0.199			
	0.020		0.199			
	0.016		0.199			
	0.016		0.199			
	0.016		0.199			
	0.023		0.203			
	0.023		0.203			
	0.023		0.203			
	0.023		0.203			
	0.025		0.185			

Table J.1. *continued*

2015	WDU (*Tektek*)		Drain (*Pengutangan*)		Δ	%
	m³/s		m³/s			
March	0.025		0.185			
	0.025		0.185			
	0.025	0.026	0.185	0.184	0.157	86%
	0.025		0.185			
	0.025		0.185			
	0.020		0.188			
	0.020		0.188			
	0.020		0.188			
	0.016		0.184			
	0.016		0.184			
	0.016		0.184			
	0.016		0.184			
	0.019		0.184			
	0.019		0.184			
	0.019		0.184			
	0.019		0.184			
	0.019		0.184			
	0.019		0.184			
	0.019		0.184			
	0.013		0.184			
	0.013		0.184			
	0.013		0.184			
	0.012		0.196			
	0.012		0.196			
	0.012		0.196			
	0.012		0.196			
	0.019		0.181			
	0.019		0.181			
	0.019		0.181			
	0.019		0.181			
	0.018		0.163			
	0.018		0.163			
	0.018		0.163			
	0.014	0.017	0.163	0.183	0.166	91%
April	0.018		0.163			

Annex K. Graphs of scenario simulations with RIBASIM

Results of the first scenario 62%

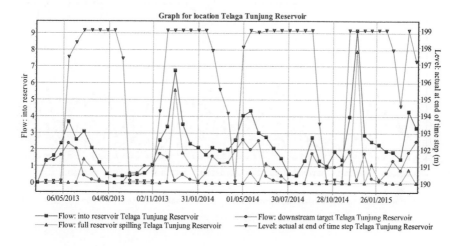

Graph for location Telaga Tunjung Reservoir

—■—Flow: into reservoir Telaga Tunjung Reservoir
—▲—Flow: full reservoir spilling Telaga Tunjung Reservoir
—●—Flow: downstream target Telaga Tunjung Reservoir
—▼—Level: actual at end of time step Telaga Tunjung Reservoir

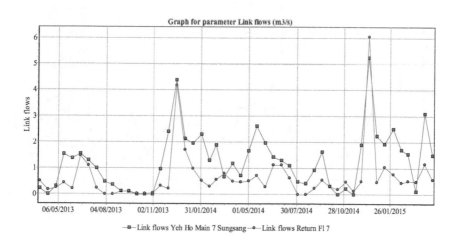

Graph for parameter Link flows (m3/s)

—■—Link flows Yeh Ho Main 7 Sungsang —●—Link flows Return Fl 7

Results of the second scenario 62%

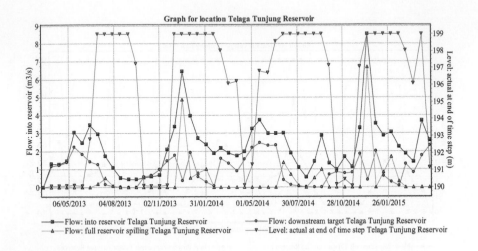

Graph for location Telaga Tunjung Reservoir

—■— Flow: into reservoir Telaga Tunjung Reservoir   —●— Flow: downstream target Telaga Tunjung Reservoir
—▲— Flow: full reservoir spilling Telaga Tunjung Reservoir   —▼— Level: actual at end of time step Telaga Tunjung Reservoir

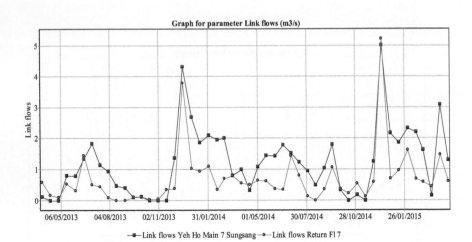

Graph for parameter Link flows (m3/s)

—■— Link flows Yeh Ho Main 7 Sungsang —●— Link flows Return Fl 7

Results of the third scenario 62%

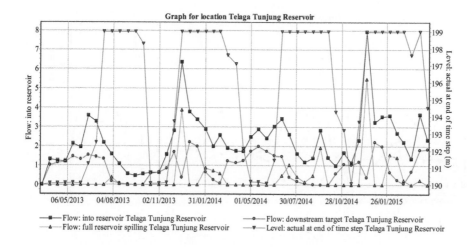

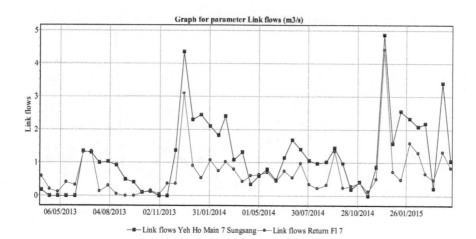

Results of the fourth scenario 62%

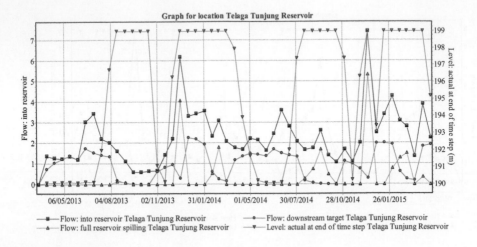

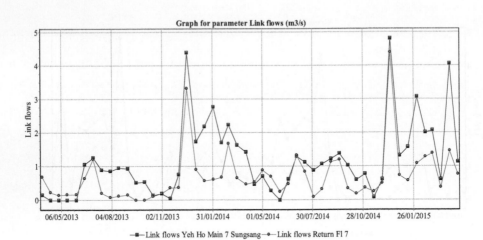

Results of the fifth scenario 62%

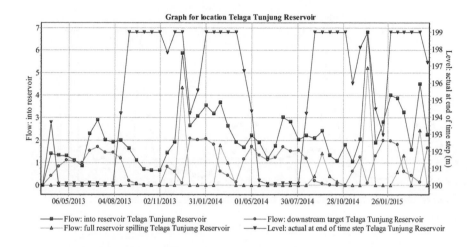

Graph for location Telaga Tunjung Reservoir

- Flow: into reservoir Telaga Tunjung Reservoir
- Flow: full reservoir spilling Telaga Tunjung Reservoir
- Flow: downstream target Telaga Tunjung Reservoir
- Level: actual at end of time step Telaga Tunjung Reservoir

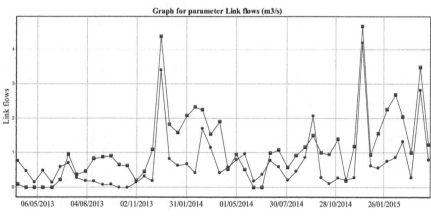

Graph for parameter Link flows (m3/s)

- Link flows Yeh Ho Main 7 Sungsang
- Link flows Return Fl 7

Results of the first scenario 21%

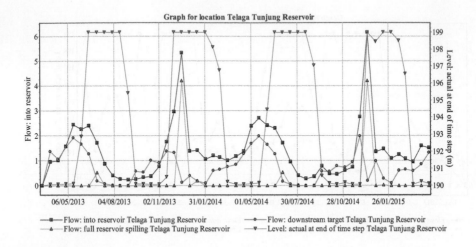

Graph for location Telaga Tunjung Reservoir

—■— Flow: into reservoir Telaga Tunjung Reservoir   —●— Flow: downstream target Telaga Tunjung Reservoir
—▲— Flow: full reservoir spilling Telaga Tunjung Reservoir   —▼— Level: actual at end of time step Telaga Tunjung Reservoir

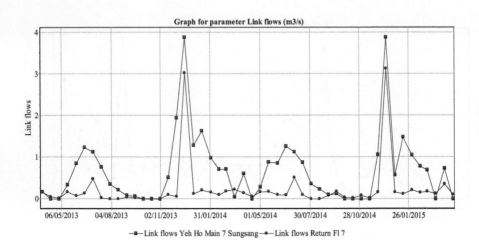

Graph for parameter Link flows (m3/s)

—■— Link flows Yeh Ho Main 7 Sungsang —●— Link flows Return Fl 7

Results of the second scenario 21%

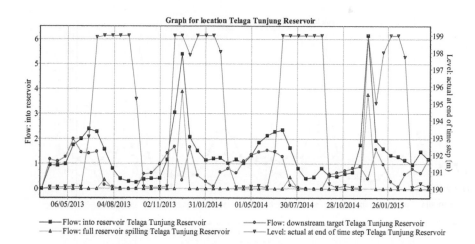

Graph for location Telaga Tunjung Reservoir

— Flow: into reservoir Telaga Tunjung Reservoir   — Flow: downstream target Telaga Tunjung Reservoir
— Flow: full reservoir spilling Telaga Tunjung Reservoir   — Level: actual at end of time step Telaga Tunjung Reservoir

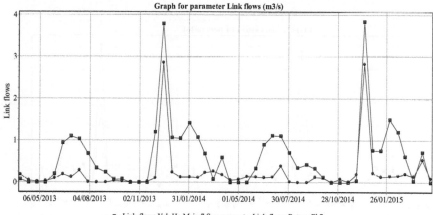

Graph for parameter Link flows (m3/s)

— Link flows Yeh Ho Main 7 Sungsang — Link flows Return Fl 7

Results of the third scenario 21%

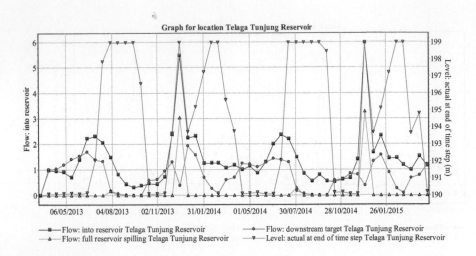

Graph for location Telaga Tunjung Reservoir

—■— Flow: into reservoir Telaga Tunjung Reservoir   —●— Flow: downstream target Telaga Tunjung Reservoir
—▲— Flow: full reservoir spilling Telaga Tunjung Reservoir   —▼— Level: actual at end of time step Telaga Tunjung Reservoir

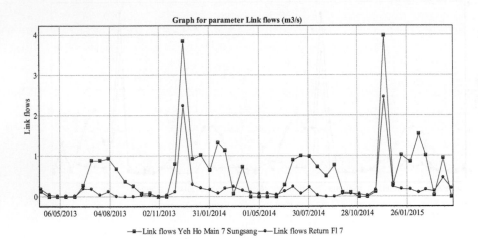

Graph for parameter Link flows (m3/s)

—■— Link flows Yeh Ho Main 7 Sungsang —●— Link flows Return Fl 7

Results of the fourth scenario 21%

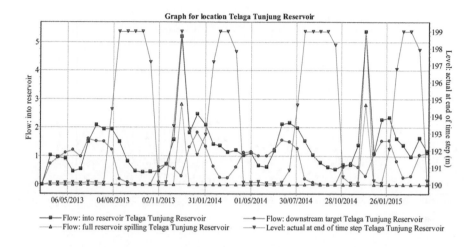

Graph for location Telaga Tunjung Reservoir

—■— Flow: into reservoir Telaga Tunjung Reservoir
—▲— Flow: full reservoir spilling Telaga Tunjung Reservoir
—●— Flow: downstream target Telaga Tunjung Reservoir
—▼— Level: actual at end of time step Telaga Tunjung Reservoir

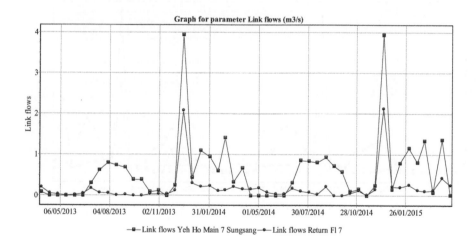

Graph for parameter Link flows (m3/s)

—■— Link flows Yeh Ho Main 7 Sungsang —●— Link flows Return Fl 7

Results of the fifth scenario 21%

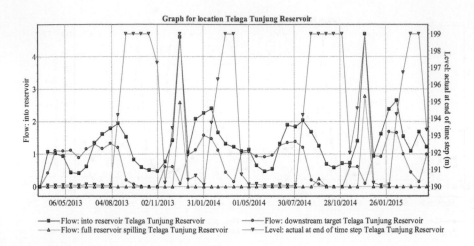

Graph for location Telaga Tunjung Reservoir

— ■ — Flow: into reservoir Telaga Tunjung Reservoir   — ● — Flow: downstream target Telaga Tunjung Reservoir
— △ — Flow: full reservoir spilling Telaga Tunjung Reservoir   — ▼ — Level: actual at end of time step Telaga Tunjung Reservoir

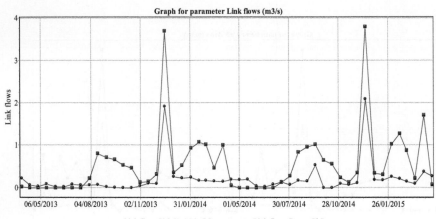

Graph for parameter Link flows (m3/s)

— ■ — Link flows Yeh Ho Main 7 Sungsang — ● — Link flows Return Fl 7

Annex L. Summary

A Subak irrigation scheme, primarily in Bali, Indonesia concerns an irrigation system of which the construction, operation and maintenance are based on agreed principles of technology, management of agriculture and a religious community. Subak irrigation schemes are an example of water resources management, distribution and supply of irrigation water in a perfect vision on social welfare in the river basin. The decision-making process in a Subak irrigation scheme takes into account political, economic, social and cultural (religious) elements. Multifunctional ecosystems in a sustainable way of agriculture are implemented in the Subak irrigation schemes, particularly in the technology of such systems. On June 29, 2012 the UNESCO World Heritage Committee formally added the Subak systems as a manifestation of the *Tri Hita Karana* philosophy to the World Heritage List.

Subak systems have been well known since the 9th Century. These systems are managed by a Subak Association based on the *Tri Hita Karana* philosophy - harmony between human beings and God, harmony between people and nature, and harmony between people and people. This philosophy underlies every activity of the Subak farmers. For managing the Subak irrigation schemes, Subak Associations and farmers pursue the Subak regulation called *Awig-awig* Subak as the togetherness consensus that in the past was announced by the King, nowadays by the Head of the Regency.

Paddy terraces in Bali create important cultural landscapes. Traditionally, the flow within a river basin has been managed using the Subak irrigation system technology. Unfortunately, this traditional technology is facing challenges: problems of water shortage and competition with other water users, that make it complicated to sustain the agriculture production of the Subak irrigation schemes. In the middle of the twentieth century, as consequence of ongoing population growth and land conversion, Balinese farmers were having difficulties in meeting the ever-growing demand for rice.

In addition, the problem of insufficient water in the dry season developed. In order to increase the irrigation water supply in Yeh Ho River Basin the Government constructed the Telaga Tunjung Dam, based on the cultivated hectares of paddy fields (*sawah*). However, since the dam was build the downstream river flow was reduced quite significantly and the process of water sharing among the Subak irrigation systems was changed. Due to this, there are serious conflicts on the water sharing among the farmers within Yeh Ho River Basin. Therefore, the main objective of this study was to develop an optimal reservoir operation strategy in relation to the operation of the Subak irrigation systems, capable to support agricultural productivity at upstream, midstream and downstream level. Detailed objectives of study were to:

- identify the type of river basin and to evaluate the land use;
- identify the contribution of the existing hydraulic structures in the main river system in supplying water to the Subak irrigation schemes;
- identify and analyze the reliability of discharge in the main river system and the inflow to the reservoir;
- analyze and determine the optimal outflow from the reservoir based on the needs of the Subak irrigation schemes within the river basin;
- simulation of the reservoir operation and operation of the related Subak irrigation systems based on the needs of the Subak irrigation schemes at upstream, midstream and downstream level to achieve optimal productivity of agriculture in a sequence time of operation related to the cropping patterns of the Subak irrigation schemes;
- to formulate recommendations on the future operation of the reservoir and related Subak irrigation systems.

Initially the study presents a literature review to portray centuries of experience with ancient Subak irrigation system management, in which Participatory Irrigation Management (PIM) is represented by three linked elements: PIM in irrigation system operation and maintenance; PIM with respect to socio-culture and economics of agriculture; PIM in light of a religious community. While there have been dynamic changes in human civilization, there are concerns how the values of the *Tri Hita Karana* philosophy can be applied consistently. The lessons learned are based on natural resources of topography, especially in the paddy terraces landscape, water resources and soils based on the principles using the *Tri Hita Karana* philosophy that influences all activities of participatory irrigation system management related to those three linked elements in Bali.

As associations for irrigation system management, Subak Associations have been already naturally adapted to Participatory Irrigation Management (PIM). The technological and socio-agricultural elements, the application of local cropping patterns and indigenous water management are regarded as the building blocks, which have been further developed in this study as scenarios of Telaga Tunjung Reservoir operation and supply to the related Subak irrigation systems. Using the Weibull formula, the historic supply data of several diversion weirs were analyzed. This analysis was able to determine the water balance of the Subak irrigation schemes behind each diversion weir. In line with this, a system approach has been applied based on the managed flows within the river basin and the characteristics of the Subak irrigation schemes.

Irrigation and drainage (irrigation-drainage) of Subak irrigation schemes concern an operational approach, which has been adapted naturally to the topography, soil and water resources of Bali. Irrigation and drainage of terrace systems became an essential element in water distribution. Subak farmers have been able to operate and manage irrigation and drainage from one scheme to another for more than a thousand years. They did also know when irrigation water reached from the highest level to the lowest level of the paddy terraces within the schemes in a river basin.

In order to sustain the agriculture production of Subak irrigation schemes there was a need for study based on farmer's perspective related to irrigation and drainage from one scheme to another. This research took place during the cultivation periods by observing the water levels at the inlet and outlet of a paddy terraces block, followed by an analysis of the trend and the amount of water being drained.

Latosol soil is the soil layer of the study region. It is one of two types of volcanic soil cover, which is the oldest of soils in Bali. The soil characteristics of Latosol can be categorized by silty clay with medium until high plasticity and brown to yellowish colours. The average value of bulk density is 0.91gr/m^3, the particle density 2.58 gr/m^3, the porosity 0.65, the average hydraulic conductivity 9.7×10^{-4} cm/hour or 2.7×10^{-9} m/s. The average root zone permeability of the paddy terraces is 75 cm/hour. The results prove that the Subak farmers irrigate for saturation of the root zone before land preparation, which requires a large amount of water.

In order to achieve an optimal scenario a technical system analysis was needed. Two common techniques have been applied, namely simulation and optimisation. The RIBASIM (RIver BAsin SIMulation) model was applied to identify the best distribution of water resources in the Yeh Ho River Basin. The focus of this study was to obtain the highest productivity related to the operation of the Telaga Tunjung Reservoir and the Subak irrigation systems with several weirs. The approach by representing the Yeh Ho River Basin with the RIBASIM model offered the possibility to find optimum results for the agriculture production. However, there was reduced availability of water for irrigation from upstream sources and from the river.

The river basin simulation uses the 80% dependable discharge, and the simulation and optimisation of the Subak schemes in Yeh Ho River Basin. While the period of land preparation is critical with respect to the water needs the scenario analysis and optimisation have been based on shifting of the starting time of land preparation (*nyorog*).

The results show that especially in the upstream and midstream of Yeh Ho River the allocation of irrigation water to the Subak irrigation schemes experienced a deficit since the water of Gembrong Spring was primarily withdrawn for domestic use, and the distribution of water from the spring was not based on applying the agreement on water sharing in practice. Therefore the first step in improved water supply would have to be to really apply the agreement about it in practice. When this will be done effective operation of the Telaga Tunjung Reservoir does not necessarily completely solve the problem of water shortage in the river basin. However, when a cropping pattern based on the fifth scenario will be applied the water deficit and yield reduction will be quite limited.

The results further show that the cropping pattern of the fifth scenario would result in an optimal overall agriculture production at 100% and feasibility of farming of 2.3 of the potential field level production for the Subak schemes. The recoverable flow considered in the RIBASIM model plays an important role in the simulation and optimisation of Yeh Ho River Basin. Two values have been applied, being 62% for the average and 21% for the minimum recoverable flow. This is supported by the *Tri Hita Karana* philosophy on harmony among people that overall could be applied in the schemes in accordance with the cropping pattern of the fifth scenario, so that agricultural productivity will be optimal. Nevertheless, under normal hydro-climatological conditions it is also possible to apply the other scenarios, especially the first scenario. This reflects the applicability of the *Tri Hita Karana* philosophy of harmony among people and harmony of people and nature. At last, it is important how Subak farmers are able to maintain harmony within the irrigation systems.

This research started quantitatively at the river basin scale and at paddy terraces block scale by composing the Yeh Ho River Basin with the Subak irrigation schemes in it. The fieldwork was done by orientation on the water sources and hydraulic structures of and within Yeh Ho River Basin. Then a closed paddy terraces block with inlets and outlets that relatively easy could be studied was analysed. These studies were based on the third element of the *Tri Hita Karana* philosophy that is the material subsystem in relation to the natural elements. This aspect implies that every paddy block of one farmer has one inlet and one outlet and that the boundaries of the Subak schemes are naturally clear. These observations were conducted during two wet and dry seasons (April 2013 - April 2015). In addition data on agriculture production within Subak Caguh upstream of Telaga Tunjung Reservoir and Subak Meliling downstream of the reservoir were collected, and soil samples were taken to determine the type and the characteristics of the soils. Based on the river basin and Subak irrigation schemes model, a scenario analysis based on the Subak irrigation systems and the *Tri Hita Karana* philosophy and Subak regulation was done. The results can be applied as a recommendation for the farmers.

Although the Subak farmers are reluctant to change to new irrigation practices, such practices can in principle be useful for them as shown in two recent studies. One of these studies concerned the application of the System of Rice Intensification (SRI) combined with intermittent irrigation (*ngenyatin*). In the farmer's perspective the SRI method is not easy applicable in practice. Nevertheless, the Subak farmers agreed that 15 days before harvesting they will dry their fields, which is one element of the SRI method.

The reservoir operation rule as applied in the fifth scenario gives the best result. The reservoir has especially an important role to increase water supply for the downstream schemes: Meliling, Gadungan and Sungsang. It also has a direct impact to sustain water conservation at the river basin scale.

Annex M. Samenvatting

Een Subak irrigatie systeem, vooral in Bali, Indonesië betreft een irrigatiesysteem, waarvan de bouw, de exploitatie en het onderhoud zijn gebaseerd op de overeengekomen beginselen van de technologie, de landbouwkundige exploitatie en een religieuze gemeenschap. Subak irrigatiesystemen zijn een voorbeeld van waterbeheer, distributie en levering van irrigatiewater in een perfecte visie op de sociale welvaart in het stroomgebied. In het besluitvormingsproces binnen een Subak irrigatiesysteem wordt rekening gehouden met politieke, economische, sociale en culturele (religieuze) elementen. Multifunctionele ecosystemen op basis van duurzame landbouw worden in de Subak irrigatie systemen gerealiseerd, met name in de technologie van deze systemen. Op 29 June, 2012 heeft het UNESCO World Heritage Comité de Subak systemen toegevoegd aan de Wereld Erfgoed Lijst als een representatie van de *Tri Hita Karana* filosofie.

Subak irrigatiesystemen zijn goed bekend sinds de 9e eeuw. Deze systemen worden beheerd door een Subak vereniging op basis van de *Tri Hita Karana* filosofie - harmonie tussen de mens en God, harmonie tussen de mens en de natuur, en harmonie tussen mensen en mensen. Deze filosofie ligt ten grondslag aan alle activiteiten van de Subak boeren. Voor het beheer van de Subak irrigatie systemen, zijn de Subak regels, genaamd *Awig-awig* Subak die zijn gebaseerd op saamhorigheid consensus, dat in het verleden werd aangekondigd door de Koning en tegenwoordig door het hoofd van de provincie, het uitgangspunt voor de Subak verenigingen en de boeren.

Rijst terrassen in Bali vertegenwoordigen belangrijke culturele landschappen. Traditioneel vindt de waterverdeling binnen een stroomgebied plaats op basis van de Subak irrigatiesysteem technologie. Helaas wordt deze traditionele technologie geconfronteerd met uitdagingen: problemen van een tekort aan water en concurrentie met andere watergebruikers, die het moeilijk maken om de landbouwproductie van de Subak irrigatiesystemen in stand te houden. In het midden van de twintigste eeuw hadden Balinese boeren, als gevolg van de aanhoudende groei van de bevolking en landconversie, problemen om te voldoen aan de steeds groeiende vraag naar rijst. Bovendien, ontwikkelde zich het probleem van onvoldoende water in de droge tijd. Met het oog om de voorziening van irrigatiewater in het stroomgebied van de Yeh Ho rivier te vergroten heeft de overheid de Telaga Tunjung dam gebouwd, op basis van de aanwezige hectaren rijstvelden (sawah). Sinds de dam is gebouwd, is de benedenstroomse rivier afvoer echter vrij aanzienlijk verminderd en is het proces van de waterverdeling onder de Subak irrigatiesystemen gewijzigd. Ten gevolge hiervan zijn er ernstige conflicten ten aanzien van de waterverdeling onder de boeren in het stroomgebied van de Yeh Ho rivier.

Daarom was het belangrijkste doel van dit onderzoek om een optimale beheer strategie voor het stuwmeer te ontwikkelen in relatie tot het beheer van de Subak irrigatiesystemen, waarmee de productiviteit van de landbouw op bovenstrooms, in het midden en benedenstrooms niveau kan worden ondersteund. Specifieke doelstellingen van de studie waren:

- de aard van het stroomgebied in kaart brengen en het landgebruik evalueren;
- in kaart brengen van de bijdrage van de kunstwerken in het hoofdsysteem van de rivier met betrekking tot het leveren van water aan de Subak irrigatie systemen;
- identificeren en analyseren van de betrouwbaarheid van de toevoer naar het hoofdsysteem van de rivier en de instroom naar het stuwmeer;
- analyseren en bepalen van de optimale afvoer vanuit het stuwmeer op basis van de behoeften van de Subak irrigatie systemen in het stroomgebied;
- simulatie van het beheer van het stuwmeer en de werking van de bijbehorende Subak irrigatie systemen op basis van de behoeften van Subak irrigatie systemen op bovenstrooms in het midden en benedenstrooms niveau om in de tijd optimale landbouw productiviteit te bereiken in relatie tot de bouwplannen van de Subak irrigatie systemen;
- formuleren van aanbevelingen ten aanzien van het toekomstige beheer van het stuwmeer en de bijbehorende Subak irrigatie systemen.

Allereerst presenteert de studie een literatuur overzicht om de eeuwenlange ervaring met het beheer van traditionele Subak irrigatie systemen te presenteren, waarbij participatief irrigatie beheer (PIM) wordt vertegenwoordigd door drie gekoppelde elementen: PIM in irrigatiesysteem beheer en onderhoud; PIM met betrekking tot sociaal culturele en economische elementen van de landbouw; PIM in het licht van een religieuze gemeenschap. Omdat er dynamische veranderingen in de samenleving zijn, bestaat er zorg hoe de waarden van de *Tri Hita Karana* filosofie consistent kunnen worden toegepast. De opgestoken lessen zijn gebaseerd op de natuurlijke hulpbronnen van topografie, in het bijzonder het landschap van de rijst terrassen, waterbronnen en de bodem op basis van de principes van de *Tri Hita Karana* filosofie die in Bali alle activiteiten van het participatieve beheer van de irrigatie systemen met betrekking tot de drie gekoppelde elementen beïnvloed.

Als een vereniging voor het beheer van irrigatie systemen, Subak verenigingen zijn al van nature gebaseerd op participatief irrigatie beheer (PIM). De technologische en sociaal landbouwkundige elementen, de toepassing van lokale bouwplannen en inheems waterbeheer worden beschouwd als bouwstenen, die verder in deze studie als scenario's voor het beheer van het Telaga Tunjung stuwmeer en de watervoorziening aan de betrokken Subak irrigaitie systemen zijn ontwikkeld. Met behulp van de Weibull formule zijn de historische watervoorziening gegevens van meerdere afleiding stuwen geanalyseerd. Op basis van deze analyse was het mogelijk om de waterbalans van de Subak irrigatie systemen achter elke afleiding stuw te bepalen. In het verlengde hiervan is een systeembenadering toegepast die is gebaseerd op de beheerde waterstromen binnen het stroomgebied en de kenmerken van de Subak irrigatie systemen. Irrigatie en drainage (irrigatie-drainage) van Subak irrigatie systemen betreft een operationele aanpak, die op natuurlijke wijze is aangepast aan de topografie, bodem en water voorraden van Bali. Irrigatie en drainage van de terras systemen is een essentieel element in de distributie van water. Subak boeren hebben irrigatie en drainage van het ene naar het andere systeem voor meer dan duizend jaar kunnen opereren en beheren. Ze wisten ook wanneer irrigatiewater van het hoogste niveau het laagste niveau van de rijst terrassen binnen de systemen in een stroomgebied zou bereiken.

Om de landbouw productie van Subak irrigatie systemen in stand te houden was er behoefte aan een studie op basis van het perspectief van de boeren met betrekking tot irrigatie en drainage van het ene naar het andere systeem. Dit onderzoek vond plaats tijdens de teelt perioden door het observeren van de waterstanden bij de inlaat en de uitlaat van een blok met rijst terrassen, gevolgd door een analyse van de trend en de hoeveelheid water die werd afgevoerd.

Latosol is de bodemlaag in het onderzoeksgebied. Het is één van de twee typen vulkanische grond en het oudste bodemtype in Bali. De bodem kenmerken van Latosol kunnen worden gekenmerkt door slibachtige klei met een gemiddelde tot hoge plasticiteit en bruine tot geelachtige kleuren. De gemiddelde waarde van het soortelijk gewicht is 0.91 gr/m^3, de deeltjesdichtheid 2,58 g/m^3, de porositeit 0,65, de gemiddelde doorlatendheid 9,7 x 10^{-4} cm/uur. De gemiddelde doorlatendheid in de wortelzone van de rijst terrassen is 75 cm/uur. De resultaten tonen aan dat de subak boeren voorafgaand aan de bewerking van het land irrigeren op verzadiging van de wortelzone, wat een grote hoeveelheid water vereist.

Om een optimaal scenario te bereiken was een technische systeem analyse nodig. Twee bekende technieken zijn in de analyse toegepast, namelijk simulatie en optimalisatie. Een model gebaseerd op generieke algoritmen is toegepast om de beste verdeling van de water voorraden in het stroomgebied van de Yeh Ho rivier te identificeren. De focus van deze studie was om de hoogste productiviteit in relatie tot het beheer van het Telaga Tunjung stuwmeer en het rivier systeem met meerdere stuwen te verkrijgen. De aanpak om het stroomgebied van de Yeh Ho rivier te modelleren met het RIBASIM model bood de mogelijkheid om optimale resultaten voor de landbouwkundige productie te vinden. Er was echter een gereduceerde beschikbaarheid van water voor irrigatie vanuit de bovenstroomse bronnen en van de rivier.

Bij de stroomgebied simulatie is gebruik gemaakt van de 80% betrouwbare afvoer, en de simulatie en optimalisatie van de Subak irrigatie systemen in het stroomgebied van de Yeh Ho rivier. Omdat de periode van bewerking van het land met betrekking tot de waterbehoefte cruciaal is, zijn de scanario analyse en de optimalisatie gebaseerd geweest op het verschuiven van de start van de bewerking van het land (*nyorog*).

De resultaten wezen uit dat er met name bovenstrooms en in het midden van de Yeh Ho rivier een tekort was bij het toewijzen van irrigatiewater aan de Subak irrigatie systemen, omdat het water van de Gembrong bron vooral als stedelijk water werd gebruikt en de verdeling van water van de bron in de praktijk niet gebaseerd was op het toepassen van de overeenkomst betreffende de waterverdeling. Daarom zou de eerste stap om tot verbetering van de watervoorziening te komen moeten zijn het in de praktijk toepassen van de overeenkomst. Als dit gebreurt leidt effectief beheer van het Telaga Tunjung stuwmeer niet noodzakelijkerwijs tot de complete oplossing van het probleem van het watertekort in het stroomgebied. Wanneer echter het bouwplan wordt gebaseerd op het vijfde scenario, dan zullen het watertekort en de opbrengst reduktie slechts beperkt zijn.

De resultaten laten verder zien dat het bouwplan van het vijfde scenario zou leiden tot een optimale landbouwkundige productie op 100% en een rentabiliteit van de landbouw van 2,3 van de potentiële productie op veldniveau voor alle Subak systemen samen. De afvoer die kan worden hergebruikt (*Natak tiyis*) speelt een belangrijke rol in het RIBASIM model bij de simulatie van het systeem van het stroomgebied van de Yeh Ho rivier. Twee waarden zijn toegepast, zijnde 62% voor de gemiddelde en 21% voor de minimale afvoer die kan worden hergebruikt. Dit wordt ook ondersteund door de *Tri Hita Karana* filosofie betreffende harmonie tussen mensen die in het algemeen op de systemen volgens het bouwplan van het vijfde scenario zou kunnen worden toegepast, zodat de landbouwproductie optimaal zal zijn.

Desalnietemin is het onder normale hydro-klimatologische omstandigheden mogelijk dat de andere scenario's, met name het eerste scenario, ook worden toegepast. Dit weerspiegelt de toepassing van de *Tri Hita Karana* filosofie betreffende de harmonie tussen de mensen en de harmonie van mens en natuur. Tenslotte is het belangrijk hoe de Subak boeren in staat zijn om de harmonie binnen de irrigatiesystemen te handhaven.

Dit onderzoek begon kwantitatief op de schaal van het stroomgebied en van de blokken met rijst terrassen door het schematiseren van het stroomgebied van de Yeh Ho rivier met daarin de Subak irrigatiesystemen. Het veldwerk is uitgevoerd op basis van een oriëntatie op de waterbronnen en op de waterbouwkundige constructies van en binnen het stroomgebied van de Yeh Ho rivier. Vervolgens is een gesloten blok met rijst terrassen met een inlaatpunt en een uitlaatpunt dat relatief gemakkelijk kon worden bestudeerd geanalyseerd. Deze studies waren gebaseerd op het derde aspect van de *Tri Hita Karana* filosofie, dat betreft het materiele subsysteem in relatie tot de natuurlijke elementen. Dit aspect houdt in dat in elk rijst blok een boer een inlaat en een uitlaat heeft en dat de begrenzing van het Subak systeem van nature duidelijk is. Deze waarnemingen zijn uitgevoerd gedurende twee natte en droge seizoenen (april 2013 - april 2015). Daarnaast zijn gegevens over de landbouwkundige productie van het stroomopwaarts van het Telaga Tunjung stuwmeer gelegen Subak Caguh systeem, en van het stroomafwaarts gelegen

Subak Meliling systeem verzameld. Ook zijn grondmonsters genomen om het type en de eigenschappen van de bodem te bepalen. Op basis van het model voor het stroomgebied en de Subak irrigatie systemen, is een scenario-analyse op basis van de Subak irrigatiesystemen en de *Tri Hita Karana* filosofie met de bijbehorende Subak regelingen uitgevoerd. De resultaten worden toegepast als een aanbeveling voor de boeren.

Hoewel de Subak boeren terughoudend zijn voor verandering naar nieuwe irrigatie toepassingen, kunnen zulke toepassingen in principe voordelig voor hen zijn, zoals twee studies hebben uitgewezen. Een van deze studies betrof de toepassing van het systeem van rijst intensificatie (SRI) in combinatie met irrigatie met tussenpozen (*ngenyatin*). Naar het oordeel van de boeren is de SRI methode in de praktijk niet makkelijk toepasbaar. Desalniettemin, zijn de Subak boeren overeengekomen dat zij hun velden 15 dagen voor de oogst droog zullen zetten, wat een van de elementen van de SRI methode is.

Het beheer van het stuwmeer zoals toegepast bij het vijfde scenario geeft de beste resultaten. Het stuwmeer speelt met name een belangrijke rol in de toename van de watervoorziening voor de benedenstroomse systemen: Meliling, Gadungan and Sungsang. Het heeft ook een direct effect op het waterbeheer op niveau van het stroomgebied.

Annex N. About the author

Mawiti Infantri Yekti was born on 12 October 1972 in Jember, East Java, Indonesia. She is the second of four children. She spent her childhood in Jember and Jakarta from 1972 until 1981. She finished her elementary school in three different places to follow her fathers profession in the Indonesian Army: in Jakarta from 1979 until 1981, in Ambon from 1981 until 1984, and in Tulungagung, when she stayed with her grandparents from 1984 until 1985. Her junior high school and senior high school were finished on 1988 and 1991 in Malang, where her parents resided after retirement. She continued her university study after she was awarded the selection of achievement student in the senior high school to enter without test at the Department of Water Resources Engineering, University of Brawijaya, Malang, East Java.

She obtained her B.Sc degree in October 1996 after writing a thesis on a productivity test of a deep tube well and an intermediate tube well for irrigation with groundwater in Pasuruan, East Java. Afterwards, she applied for a job and a master scholarship of the Indonesian Government (DUE Project) in Februay 1997. In the same year, she got a job as assistant engineer in the Project Type Sector Loan (PTSL) Nippon Koei Ltd with the local consultant PT Tata Guna Patria. She was only from July until August in this job with the task of reviewing of water engineering projects. Owing to this she was awarded a DUE scholarship. She started to study for a master degree in September 1997 and she chosed the subject of water resources at Graduate Studies of Civil Engineering of Universitas Gadjah Mada (UGM), Yogyakarta. The DUE Project was an Indonesian Government Scholarship Programme for selecting the new lecturers for several state universities during 1997 - 1999.

She obtained her master degree in May 2000 after writing her thesis on *Optimization of reservoir operation analysis*. In April 1999 she had already officially become a civil servant. In September 2000 she started to lecture at the Department of Civil Engineering, Udayana University, Bali. She married to Ristono in July 2001. They have a son, Narayana Radya Aydin, who was born in December 2002, and a doughter, Larasati Ridha Alisa, who was born in January 2005.

She teaches several subjects related to water science engineering, such as irrigation, fluid mechanics, hydraulics and hydrology for undergraduate students. She is also supervisor for undergraduate final projects and practical projects and academic supervisor of more than 30 undergraduate students. She had 21 topics of research with her undergraduate students.

In 2003, she started as assistant lecturer to teach for graduate students on subjects as advanced hydrology, earthfill dams and groundwater movement. In this period, she felt that she had limited experience with the applications in water science engineering to teach in the graduate program since she graduated from bachelor and master studies. Then in 2004 she started to be involved as an expert. Since then she has had 21 involvements in water engineering design project.

With respect to organization experiences, she was responsible for two divisions in the Department of Civil Engineering in Udayana University, from 2004 until 2010 she was head of the reading room (small library), and from 2004 until 2008 secretary in the laboratory of hydrology and hydraulics. Moreover, she is member of three professional organizations: Indonesian National Committee on Large Dams (INACOLD), Indonesian Commission on Irrigation and Drainage (INACID), Indonesian Association of Hydraulic Engineers (HATHI). She has an expertise certificate of the Indonesian Society of Civil and Structural Engineers (HAKI) in the specialization of water resources engineering.

Accepted papers by international journals:

- Yekti, M. I., Schultz, B., Norken, I. N., and Hayde, L. Discharge analysis for a system approach to river basin development with Subak irrigation schemes as a culture heritage in Bali. *Agricultural Engineering International: CIGR Journal.* http://www.cigrjournal.org/index.php/Ejounral/index. Accepted;

- Yekti, M. I., Schultz, B., Norken, I. N., and Hayde, L. Learning from experiences of ancient Subak schemes for participatory irrigation system management in Bali. *Irrigation and Drainage.* http://onlinelibrary.wiley.com/journal/10.1002/(ISSN)1531-0361. Accepted.

She has 15 publications in international/national proceedings and national journals:

- Yekti, M. I., Schultz, B., Norken, I. N., Gany, A. H. A. and Hayde, L. 2014. Irrigation-drainage of Subak irrigation schemes: a farmer's perspective over a thousand years. Proceeding 12th ICID International Drainage Workshop, Drainage on Water Logged Agricultural Areas, 23 - 26 June 2014, St. Petersburg, Russia;

- Yekti, M.I., Norken, I.N., Schultz, B., and Hayde, L., 2014. A role concept of reservoir operation for sustainable water supply to Subak irrigation schemes: case study of Yeh Ho River Basin. Proceeding International Symposium on Dams in global environmental challenges. The 82th Annual Meeting of ICOLD (International Commission on Large Dam), 1 – 6 June, 2014, Bali, Indonesia;

- Yekti, M.I., Gany, A.H.A., and Schultz, B., 2014. Learning from decades of experience with Subak ancient Participatory Irrigation Management in Bali. Lecture note at Water and Land Management Institute (WALMI), 21 - 24 January 2014, Aurangabad, Maharashtra, India;

- Yekti, M.I., Schultz, B., and Hayde, L., 2013. Development of a conceptual approach to manage flow of Subak irrigation schemes in Bali, Indonesia. Proceeding 1st World Irrigation Forum, 29 September - 3 October 2013, Mardin, Turkey;

- Yekti, M.I., Schultz, B., and Hayde, L., 2012. Challenge of runoff regulation to supply paddy terraces in Subak irrigation schemes. Proceeding 7th Asian Regional Conference of International Commission on Irrigation and Drainage (ICID), 24 - 28 June 2012, Adelaide, Australia;

- Siladarma, I G.B., and Yekti, M.I., 2010. Simulation of thermal water dispersion from cooling water system of Kubu Steamed Power Plant, Karangasem Region. Proceeding IATPI National Seminar VI, 29 July 2010, Bali, Indonesia;

- Siladarma, I G.B., and Yekti, M.I., 2009. Predication of thermal water dispersion from Kubu Steam Power Plant - No. ICOE 273. Proceeding International Conference in Ocean Engineering (ICOE), 1 - 5 February 2009, IIT Madras, Chennai, India;

- Yekti, M.I., 2008. Water surface degradation of Lake Buyan using water balance analysis. Proceeding HATHI Seminar, 21 – 23 August 2008, Palembang, South Sumatera, Indonesia;

- Siladarma, I G.B., Yekti, M.I., and Permana, G. I., 2007. Impact of land use changing on design flood at sub river basin of Ayung, Journal of Water, BITK Vol. 13 No. 3 July 2007, accreditation 23a/DIKTI/KEP/2004, ISSN 0854-4549, Department of Civil Engineering, Engineering Faculty of Diponegoro University, Semarang, Indonesia;

- Yekti, M. I. and Kurniawan, P. A., 2007. Relation height metacentrum and draft for three types of pontoon. Research Report the Directorate General of Higher Education, Ministry of Education, Number 010/SP2H/PP/DP2M/III/2007;

- Yekti, M.I., Arsana, I G.N.K., and Luthfi, 2006. Evaluation reservoir capacity of Telaga Tunjung with three critical period method: behavior analysis, dincer, sequent peak algorithm. Journal of Water, No. 1 Years 13 July 2006, accreditation 23a/DIKTI/KEP/2004, ISSN 0854-4549, Laboratorium Water Flows, Department of Civil Engineering, Engineering Faculty of Diponegoro University, Semarang;

- Yekti, M. I., 2004. Productivity test of deep tube well and intermediate tube well for irrigation groundwater area, Journal of Water, No. 1 Years 11th July 2004, accreditation 23a/DIKTI/KEP/2004, ISSN 0854-4549, Laboratorium Water Flows, Department of Civil Engineering, Engineering Faculty of Diponegoro University, Semarang;

- Yekti, M.I., 2004, Fluids mechanics lecture book, Department of Civil Engineering, Engineering Faculty, Udayana University, Bali;

- Yekti, M.I., 2001. Optimization analysis of reservoir operation, Journal of Civil Engineering Vol.5 No. 9 July 2001, ISSN No. 1411-1292, Udayana University;

- Yekti, M.I., 2001. Precipitation - runoff model using Mock method for Sermo Reservoir. Paper note of Department of Civil Engineering, Engineering Faculty, Udayana University, Bali.

Netherlands Research School for the
Socio-Economic and Natural Sciences of the Environment

D I P L O M A

For specialised PhD training

The Netherlands Research School for the
Socio-Economic and Natural Sciences of the Environment
(SENSE) declares that

Mawiti Infantri Yekti

born on 12 October 1972 in Jember, Indonesia

has successfully fulfilled all requirements of the
Educational Programme of SENSE

Delft, 1 June 2017

the Chairman of the SENSE board

Prof. dr. Huub Rijnaarts

the SENSE Director of Education

Dr. Ad van Dommelen

The SENSE Research School declares that Ms Mawiti Infantri Yekti has successfully fulfilled all requirements of the Educational PhD Programme of SENSE with a work load of 38 EC, including the following activities:

SENSE PhD Courses

o Environmental research in context (2012)
o Research in context activity: 'Creating video impression of research on: Challenge of Runoff Regulation to Supply Paddy Terraces in Subak Irrigation Schemes' (2012)
o SENSE Writing Week (2013)

Other PhD and Advanced MSc Courses

o Service Oriented Management Irrigation System, UNESCO-IHE, Delft (2011)
o Hydro-informatics for Decision Support, UNESCO-IHE, Delft (2011)
o Academic Writing, UNESCO-IHE, Delft (2012)
o Summer Course on Morphological Modelling using Delft3D, UNESCO-IHE, Delft (2013)

Management and Didactic Skills Training

o Member of the commission of assessment of the public policy of HATHI (Indonesian Association of Hydraulic Engineers) Province Bali (2014)

Oral Presentations

o *Challenge of runoff regulation to supply paddy terraces in Subak irrigation schemes.* Irrigation Australia 2012 Conference, ICID, 24-30 June 2012, Adelaide, Australia
o *Development of a conceptual approach to manage flow of Subak irrigation schemes in Bali, Indonesia.* First World Irrigation Forum, 29 September - 5 October 2013, Mardin, Turkey
o *A role concept of reservoir operation for sustainable water supply to Subak irrigation schemes. Case study of Yeh Ho River Basin.* ICOLD (International Commission on Large Dams) 82nd Annual Meeting: Dams in Global Environmental Challenges, 2-6 June 2014, Bali, Indonesia
o *Irrigation-drainage of Subak irrigation schemes: a farmer's perspective over a thousand years.* 12th ICID International Drainage Workshop: Drainage on Waterlogged Agricultural Areas, 23-26 June 2014, St Petersburg, Russia

SENSE Coordinator PhD Education

Dr. ing. Monique Gulickx

Printed and bound by CPI Group (UK) Ltd, Croydon, CR0 4YY

18/10/2024

01776210-0002